Sportscasting

ELECTRONIC MEDIA GUIDES

Elizabeth Czech-Beckerman *Managing Electronic Media*

Ken Dancyger *Broadcast Writing: Dramas, Comedies, and Documentaries*

Donna L. Halper *Full-Service Radio: Programming for the Community*

Donna L. Halper *Radio Music Directing*

William Hawes *Television Performing: News and Information*

Robert L. Hilliard *The Federal Communications Commission: A Primer*

John R. Hitchcock *Sportscasting*

Andrew F. Inglis *Satellite Technology : An Introduction*

Carla B. Johnston *Election Coverage: Blueprint for Broadcasters*

Marilyn J. Matelski *TV News Ethics*

Marilyn J. Matelski *Daytime Television Programming*

Robert B. Musburger *Electronic News Gathering: A Guide to ENG*

Sportscasting

John R. Hitchcock
Vincennes University

Focal Press
Boston London

Focal Press is an imprint of Butterworth–Heinemann.

Recognizing the importance of preserving what has been written, it is the policy of Butterworth–Heinemann to have the books it publishes printed on acid-free paper, and we exert our best efforts to that end.

Library of Congress Cataloging-in-Publication Data
Hitchcock, John R. (John Robert)
Sportscasting / by John R. Hitchcock.
p. cm. —(Electronic media guide)
Includes bibliographical references.
ISBN 0-240-80062-1 (paperback)
1. Mass media and sports—Vocational guidance. 2. Television broadcasting of sports—Vocational guidance. 3. Radio broadcasting of sports—Vocational guidance. I. Title. II. Series.
GV742.H58 1991

070.4'49796023—dc20 90-40740
CIP

British Library Cataloguing in Publication Data
Hitchcock, John R.
Sportscasting—(Electronic Media Guide)
1. Radio programmes. Television Programmes. Radio and television.
Sports programmes. Commentating
I. Title II. Series
070.449796

ISBN 0-240-80062-1

Butterworth–Heinemann
80 Montvale Avenue
Stoneham, MA 02180

10 9 8 7 6 5 4 3 2 1

Printed in the United States of America

Contents

Preface

There are numerous career opportunities in the sports and media world. Many entry-level positions are in radio; the main thrust of this book is aimed accordingly. Television sportscasting jobs are more glamorous and usually pay better, but are much harder to obtain. The basic skills that are introduced in this book may be applied to a variety of related sports and media fields.

The materials and techniques found in this book are based on practical experience and not just theories of what *could* or *should* happen. This book examines the basic information needed to get started in a variety of sportscasting careers. The techniques and situations covered here must be adapted to each situation encountered. Every radio and television station operates in its own way. Be flexible enough to meet the requirements of each situation. This material is not carved in stone and should be used as a springboard into each situation.

In sportscasting, or any career area, it's the intangibles that make the difference. Personality, audience appeal, diligence, loyalty, integrity, willingness to work, and the ability to learn are all factors in a successful career. A vital area is your attitude. Be willing to go the extra mile and never be satisfied with a "that's good enough" approach.

1 Sportscasting Jobs

Becoming a sportscaster may not be as easy as it seems. Sportscasting is a fun, exciting, and rewarding job. Because it is a glamour job, the competition for the limited number of positions is intense. Athletes and coaches are trading in on their name value with the public and using their expertise to take over many jobs that formerly had been held by professional announcers.

There are excellent reasons for becoming a sportscaster. There is the glamour that goes with the job and the audience recognition that goes with it. It is a great way to become part of the sports scene, even if you aren't a participant. It is a big ego boost to be recognized and rewarded for your efforts. There is money to be made and a chance to travel and meet the "stars" of the game. You usually get good seats to work from and are treated with respect. You meet some nice people who may be of great assistance to you.

There are also drawbacks to this job. It takes long hours, usually at low pay, to get the job done right. It is a tension-packed job with intense pressure to have everything ready to go to get the game on the air. Once you do get the game on the air, you must perform before a critical audience. You may run into problems and hassles with people and organizations. You may work in noisy, crowded conditions with major distractions and problems. Most college and high school games are conducted on the weekends and at night. You may have to work holidays to cover events. It takes plenty of work to prepare to do a game or event correctly. You can't just step up to the microphone and start talking. You must prepare emotionally, mentally, and physically to give the audience a good performance. It can be physically tough to have a good game if you are tired, not feeling well, or have poor facilities to work from. The higher up you go on the sportscasting ladder, the better the conditions and support you receive. Your performance must also be that much better to keep the audience happy. You must become a sports expert. It is very time-consuming to do the job right.

REPORTERS AND ANCHORS' SKILLS

While play-by-play (PBP) and color commentator positions tend to be the more glamorous and fun work in sportscasting, reporter and anchor jobs are more common. Both radio and television use reporters and anchors for basic sports coverage. There is a limited amount of time to entertain the audience with the sporting news and scores. You have only 2 to 5 minutes to get the job done. You must be concise, entertaining, and as thorough as possible in the time allotted. In television, you must pick the appropriate video to entertain your audience; in radio, you pick the voices and sounds that supplement your reports. The key word is always *time*. Time determines the materials that are used in your sportscast.

Know Your Audience

The entire sportscast must be centered around the wants and needs of the audience. Each station has a desired *target* audience it wants to attract. A good sportscaster will know who the audience is and what it wants to know. The delivery and style must fit the audience, station image, format, and the sportscaster's personality. The materials that are chosen and how they are delivered to the audience are key in building the desired image with the audience. You must inform, explain, and entertain the audience in a style that leaves a favorable impression with them so they will keep coming back for more. The pace, power, intensity, emotional coloration, and personal style must combine with the prepared materials to obtain the image you want the audience to have of you. In sportscasting, you must combine the importance of news with the flash and excitement of entertainment. You must be confident and comfortable with the materials and situations you are dealing with. You must appear to be the expert everyone in the audience knows that they are. If you appear uncertain or hesitant, you may lose your credibility with the audience full of "experts."

One of the key words in radio and television sports is *local*. While the audience wants to know the scores and information on the national scene, the local connection can be vital. If a local product is playing for a professional or major college team, there will be local interest about that person and team. Keeping up on the local scene and the local connection to the bigger sports scene is vital in keeping the interest of the audience.

KNOW YOUR CONTENT

A reporter at the scene of a game or event must keep in mind the needs of the sportscast. You must understand the context the material will be put into and the materials that may surround it. Reports from the scene must be short, concise, and entertaining to the audience. Being on the scene to file a report can add to the immediacy and credibility of the report. It can also create problems due to noise, lack of noise, or the situation when the report is being filed. Write out as much information as possible and practice delivering the material prior to the recording or transmission to the station. At a minimum, write down the main items you want to use, in the order you want to cover them. Then focus your thoughts from one item to the next on your list. The amount of time available when you are reporting is important. You must cover the high points and keep the story interesting. You must know if it will be going live to the audience or be recorded for later broadcast. If you make a mistake, can you do it over or do you have to live with it? The pressure of time is the reporter's constant companion.

The Call-In Show

A popular form of sports on radio is the sports talk/interview or call-in show, a great platform for the non-play-by-play sports audience and advertisers. Many stations are finding that this type of show attracts a valuable audience. The format may include live or recorded interviews, telephone call-ins from the audience, or a combination of both. The audience call-in format gets the audience involved, which is valuable to the station and advertisers.

When you open the phone lines, the audience expects to be entertained and informed. You must be, or learn to be, an expert on many sporting topics. You will need to find a quick way to get the information needed for the audience. Reference books and magazines, files, and computers are all used to get the information needed. Athletes and coaches base their opinions on their expertise.

Each show is like a new opening night. Each show must contain enthusiasm and vitality for the audience. There must be enough variety, entertainment, and enthusiasm to catch and hold the intended audience. An attention-holding pace must be found and maintained or the show will drag and get boring. You must be expert enough to point out significant facts and translate information for the audience to understand. There will be situations that require you to ad-lib or "make do" until you can move on. Learn how to answer a question with a question in an entertaining way. While you are the mainstay of the show, the guests you put on the air are very important. How much show biz or information you use depends on your personality, style of your show, and what the audience expects. Some hosts deliberately antagonize their audience to provoke a response. Find out what works for you.

The Analyst-Expert

Many radio and television stations have regularly scheduled coaches' shows during the sports season. The coach or coaches usually come to the studio and talk about the games and the season. Television stations will show videotapes of the game action while the coach and moderator discuss each situation. These shows are formatted to fit the time frame, style of the moderator, availability of audio or video cuts, and the type of coach. The moderator needs a list of questions or areas of interest that can be used on the show. Television may use the teleprompter system to show information on a screen over the camera lens, or a blackboard may be set off-camera to cue the moderator.

A job usually held by a *sports expert*, such as a coach or athlete, is that of color commentator or analyst, whose major contribution to the game is pointing out special pieces of information that add to the knowledge and enjoyment of the audience. In some cases, the analyst is there for personality. The analyst's job includes looking for the drama of the game and helping the play-by-play announcer set the scene and build on it. The key moments in the game should be noted and commented on or interpreted through the expert's eyes.

Having two announcers is important in football and baseball coverage due to the stop-and-go nature of the sports. Basketball doesn't have much time for comments between plays since the action is basically nonstop. The analyst is more valuable in television coverage because the picture carries the action instead of the announcer's voice. Television replays give analysts plenty of opportunities to use their expertise. Three people doing a game can be very confusing unless the announcing is strictly controlled and formatted accordingly. If all three can work together smoothly, it does add extra depth to the broadcast. The three-person approach is more practical on television than on radio. Most stations will use a two-person team to handle the remote broadcast.

The Engineer

Some radio stations send along an engineer to handle all the technical details and to mix the audio signals at the game. This takes a major concern away from the announcers and allows the engineer to concentrate on maintaining the technical quality of the broadcast.

The Statistician

Another job is that of the statistician. If the statistician is able to keep the numbers for the announcers, the announcers can concentrate on the game action. A complete set of statistics gives the audience a better, or at least different, perspective on the game. Sports such as football and baseball have many situations that allow for statistical analysis. Open time frames or spaces can be filled with this information. A well-kept set of statistics gives the broadcast greater credibility with the audience and gives the announcers material that can be used to explain the actions and trends in the game.

There are many opportune places to give the audience a statistical view of the game, such as at the quarter, halftime, or inning break, end of game wrap-up and reports. Try not to get carried away with the numbers. It is easy to bore your audience with too many figures. Be selective in the statistics you use so they add to the audience's understanding of the game. Many games and event officials will provide statistics for your use during and after the action. It is a good idea to keep your own statistics as a backup and as a check on the statistics you receive. In some cases, the statistics aren't available to be used at halftime or at the end of the game when you are wrapping up.

The Spotter

The spotter's job is to locate and identify the players involved in the action. The spotter is used mainly in football coverage, due to the number of players involved, the complexity of the game, and the distance from the action the announcers are located. The spotter (or spotters) concentrates on the action to locate the people directly involved or those considered important in the play. If a gang tackle is made, a fumble occurs, or several people are involved in a quarterback sack, the spotter can identify the players involved and what their roles were in the play. The spotter is also responsible for identifying new players or any changes in personnel. The spotters may be seated or stand beside or between the announcers to point to a chart or list of names to show the players involved. The spotter may whisper off-mike to the announcer information to be passed along to the audience. A separate audio channel to the announcer's headset may also be used to provide information.

While it is nice (but in many cases considered a luxury) to have extra help on a remote broadcast, it may not be practical or possible. Press box facilities may be too small to allow for any extra personnel. The sponsoring organization may allow only the two-person announcing staff to operate. It may be difficult to locate, train, and depend on extra help. The most common reason is that management doesn't want or cannot afford to pay for extra personnel. For the bigger, more profitable events, all three areas are filled and a five- or six-person crew is used to get the best possible results. An extra announcer may go along to work the sidelines and collect information and interviews. A producer-director may be included to handle the overall control and sound of the radio broadcast.

2

Sports Interviewing and Reporting

Every sports reporter, anchor, or show host must know how to do a successful interview. The purpose of the interview determines the technique and style to be used. How you prepare for the interview depends on the type of story or information you are seeking. Each interview should be handled according to the situation. How you approach a coach in a pregame interview will be different from a postgame interview after a tough or controversial loss.

SETTING UP THE INTERVIEW

Setting up an interview may be difficult. Some people don't want to have anything to do with a reporter. Sometimes, just getting to the source is difficult. If secretaries turn you away, call early, at lunch time, or late in the day to try to avoid them. Some reporters will interview associates of the target, hoping that news of the interview will filter back to the desired person. Other reporters will wait at the person's home or office and "ambush" them when they arrive. Many people who find themselves in the news are glad to talk with a reporter so they can state their case, counter a charge, or gain some favorable publicity. Others, whose portrayal may not be flattering, may be more reluctant. Try to persuade the source that the interview may help them. Explanations are often less damaging than insinuations or rumors.

INTERVIEWING STRATEGIES

There are many strategies that can be employed during an interview. A good interviewer knows the art of small talk, of making conversation. It is important to get the subject into a talking frame of mind. Start with easy, comfortable questions to get them going. Sources who think the reporter knows about the subject are more likely to speak frankly and fully instead of feeling they must explain all the details. If they appear skeptical, open with a question that shows you've done your homework and have knowledge of the subject and situation. It might pay to be low-key and let them relax. Try to be sympathetic and understanding. If the source is tense or nervous, it may be beneficial to increase their anxiety level. People tend to make mistakes or say more than they intend to when under pressure. A leading or rhetorical question can be useful if the subject is hostile, shy, or feels intimidated. Sense when *not* to ask a question. By remaining silent, the subject may restate the answer in a clearer way or cover a different area. Try not to step on their answers with your question or comments.

Keep the Interviewee Talking

Sports reporters usually work to obtain "short essay" responses. A person being interviewed should be doing most of the talking. If something goes wrong, a reporter cannot be afraid to ask for a repeat of a statement or a question. Most reporters have an idea of what they need from the person being interviewed and will try to guide the answers in the direction they want. Most interviews are designed to obtain specific pieces of a story so the information flow is tightly controlled. Don't let the subject ramble on or change the subject. It may be appropriate for the reporter to suggest that any comments will be more effective if they are short and to the point. Doing a good interview requires firmness and friendliness in various combinations. A reporter is the representative of the audience and should pose questions that the audience would want answered. Listen carefully to what the subject is saying. Don't anticipate or assume what the answer will be, even if it's a familiar topic.

When a reporter talks to a possible source of information, the source may ask that the interview be "off the record." If the reporter agrees, the information provided cannot be used except to lead to other information, or to confirm information from a different source. No mention of where the information came from can be used. Other sources may say they will give information only for "attribution," which means they will allow the reporter to use the information, but not to reveal the actual name of the person. This method may result in a story attributed to a "source high in the administration." If the source agrees to an "on the record" interview, which is generally assumed if no objection is raised, anything they say can be used and quoted. Some sources will provide background information to help the reporter grasp the extent and meaning of a situation or story. Reporters may also handle "leaks" or "trial balloons." A leak is when a source shares sensitive information with one reporter. The reporter must weigh the value of an exclusive story against the reliability of the source and if the accuracy of the information can be verified. A trial balloon is information given to a reporter to test the public reaction to an idea before it is formally announced to the public.

Questions

The way an interview question is structured determines the type of answer. An *open-ended* question doesn't specify an answer, so subjects must use their own words and thoughts in response. This style appears less threatening and allows for some flexibility. A *closed-ended* question requires specific answers from which to choose. This type of question is good for obtaining details and specific information.

During an interview, avoid leading questions and questions with "yes" or "no" answers. Avoid overly complex questions that lead to confusion by both the guest and the audience. No matter what the situation, you must stay calm and in control of the interview. You may have to be persistent, and possibly a bit pushy, and not put off by evasive answers or no answer at all. Time is always a factor during an on-scene interview. How much time the person has available for the interview and how much material you need is important. Does the interview go on the air as is or will you have time to edit the material? If it is a live report or a complete interview, work up an open and a close to give it that polished professional sound.

If you are gathering materials on the scene, let them tell the story, then go back and fill in any information gaps with your questions. If you have a good overview of the situation, you can guide the remaining questions into areas of value. Ask questions that require some explanation. Use those explanations for good quotes or statements. Constantly work to force the subject to illustrate ideas with concrete examples or stories. Make sure you understand the answer. Don't let someone hide behind an unclear answer. Use follow-up questions to translate jargon, clear up any confusion, respond to evasions, or simply to get a better quote.

If you are trying to get a source to admit or say something he doesn't want to say, there are several strategies and techniques that can be tried. Use questions that force the subject into a corner and place on the record general statements that make it more difficult to evade specific questions. It may work to spread trickier questions throughout the interview to surprise the person or lead them into a contradiction. Another method is to ask a series of tough questions, leaving little room for evasion. Try to put your questions in an illogical order before the subject figures out what information you are trying to obtain. Ask questions for which you know the answers and check his answers against yours. This may help force some answers when your source is confronted with your knowledge of those incorrect answers. Try to be casual when popping the "big" question. Sometimes you can bluff answers from a subject if you have enough facts on which to base the bluff. Another method is to deliver short questions, one at a time as in a courtroom, so each query is on a direct point and each answer leads to the next question. Make the source answer each question with specifics and continue until you get the answer or the interview is finished. Most interviewers will hold the "big" questions until near the end of the interview to avoid an early cutoff of information.

If you are to do an in-studio interview, preparation is a key to success. You should know about 60% of the answers the guest will give before the interview. Make sure the guest is comfortable with the situation after arrival. Do not submit specific questions to the guest before the interview. Make sure the general areas you want to talk about are clear so your guest can be mentally prepared. This will avoid canned or rehearsed-sounding answers. Limit the topic to be covered to the time you have allotted for it. Establish the credentials of the guest and the importance of the topic at the beginning of the show. Write out your introduction and closing lines for the interview. This will get you off to a good start and let you finish with style, avoiding fumbling around. You want to leave the guest and the audience with a good impression.

The guest is the star of the interview. Listen attentively to the replies to your questions. Based on the answers, new questions may occur to you in areas that should be explored or followed up. If you don't appear to be paying attention to the answers, it will be hard for the guest to stay interested in the conversation. Guide the interviews by the wording and selection of the questions. Keep your questions short and to the point, but don't be rude about it. Your audience will prefer anecdotes and stories to receiving only facts and figures. Write out more questions than you will need prior to the interview. This helps avoid any gaps in the interview and assures that you cover the material you want covered. During the interview, write down questions or memory joggers for later use or reference. Always be ready with the next question to keep the interview moving at an entertaining pace. If it gets slow, it gets dull. Point out and underscore important points and answers for the benefit of the audience. Ask questions based on what the audience wants to know. Work your way up to a controversial or critical question with care. Establish the ground work before you hit the tough questions. If the interview is with more than one person, interview them one at a time to avoid confusion and splitting the audience attention. At the end of the interview, thank them briefly and close as you have already planned.

VERBAL AND NONVERBAL CUES

It is important to understand the verbal and nonverbal communication that takes place during an interview. What a person says and how it is said is important to the meaning of the information. Examine the choice of words, how the words are put together and in what order. Listen carefully to the subject's voice. Note the delivery, pace, inflection, amount of intensity, volume, tone, pitch, and emotional content. Voices can carry more than one meaning, so the interviewer must listen for fine nuances or meanings. Examine the content of the answers and look for any patterns or trends. Look for anything that doesn't seem to fit or strikes a wrong note and figure out why. If you are familiar with the person, it is easier to "read" the meaning or lack of meaning by comparing that performance to previous meetings.

In addition to the verbal clues, nonverbal clues such as body language and facial expressions may be indicators of meaning. A careful watcher should examine the subject's posture. Is the guest standing tall or slumped over? The person standing tall will appear more confident and in control, while the slumped posture appears more indifferent or uncaring. These are generalities, but may provide clues to the meaning of a statement. Good interviewers instinctively read their subjects. Look for head movements, lifting eyebrows, hand or arm gestures, body stance, muscle tenseness. Watch how long a position is held and how often it shifts. Different positions can often be related to the emotional state of the person. The subject's eyes can transmit subtle nuances that the careful observer will note. The eyeball itself doesn't show much (even though there is an unconscious widening of the pupil if the eye sees something pleasant), but the directness and length of a gaze, opening of the eyelids, squinting, or dozens of little manipulations of the skin around the eyes can transmit many meanings. If the subject looks away during the conversation, it may betray lack of confidence or mean something is being concealed.

Be very careful if you decide to tie a specific posture or change of movement to a specific vocal statement. A shift in posture means that something is happening, but it doesn't tell us what. Any verbal or nonverbal clues must be put into the context of the situation. The nonverbal response you get may be false. Cultural differences can play a large part in body movement. Latin cultures tend to use more arm and hand gestures than northern European cultures. Aside from cultural differences, the subject may deliberately use gestures or motions designed to project a desired meaning. Politicians, actors and others in public view know how to project a message with nonverbal communication that suits their needs. A reporter-interviewer can use body language and verbal approach to encourage answers.

A good sports reporter gets a story by digging, being industrious, aggressive, and asking questions. Reporting is a search for answers that are usually uncovered by intensive questioning. Reporters need to use questions that are to the point and correctly phrased to get a desired response. The questions might cause discomfort. A reporter must be firm and willing to press on. Tough or unfriendly questions have a way of ending the flow of information, so those questions are usually left until near the end of the session. Once a story angle appears, questions are aimed at getting facts that fit the story angle. A sports reporter needs enough background information to put the story into perspective and to select the proper information. Finding the story angle (something different or spicy) may require close attention to seemingly insignificant details. A reporter's final question to a subject should be, "Is there anything else I should know about this?"

TAPE OR AUDIO CUTS

An important area for any radio sports reporter is to find and use audio recordings, generally referred to as *tape* or *audio cuts.* There is a variety of cuts used in sports reports. An *actuality* is the voice or sounds of a newsmaker. Actualities allow the audience to hear comments or sounds from the source instead of a reporter's version of those comments. Actualities may add colorful details, emotional impact, or to the informational content of the story. Actualities may provide eyewitness reports, subjective comments, or expert knowledge about a subject. Any natural sounds of a newsworthy event that are recorded for use are called *nat* [natural] *sounds*, *in the clear*, or *wild tracks.* The length depends on its importance or significance to a story. An actuality cannot stand alone and needs supporting information, usually provided by a lead and a tag. The *lead* and *tag* are scripted information read before (lead) and after (tag) the actuality is played for the audience. The audio cut must add to the story, be of good quality, and make sense.

Another type of tape is the *voicer.* A voicer is a recorded report from a reporter at the scene. The reporter's monologue may describe the scene as it occurs. This type of voicer is called a *ROSR* (pronounced RO-zer) and stands for "radio on-scene report." The reporter can pretend it is a live broadcast to help capture the color or feel of an event. It can be edited to fit the time and story back at the studio. Voicers will usually stand on their own and may or may not need additional material for audience understanding.

A *wrap* (wraparound) is another type of audio cut. The wrap is a self-contained open and close report wrapped around an actuality. The reporter sets the scene and provides the information, plays the actuality, and then provides additional information. A wrap can be played in its entirety and should not require other support materials in the sportscast.

Another type of audio cut is the *Q & A* (Question and Answer). Q & A's are a kind of actuality that includes the reporter's voice as well as the newsmaker's voice. The Q & A form is used when you cannot edit out the question, the question in its own right is newsworthy, or is needed to understand the answer. It may also hold the interest of the audience longer than an actuality since there are two voices in conversation.

Audio cuts can be edited to fit the time and situation as long as the speaker's meaning or intent is not distorted. It is usually best to leave out changeable facts, so the report won't become dated or wrong. It is usually best to leave the introduction or lead-in and the beginning of the voicer as if the scene has already been set by the anchor. Often the reports will start with the word *the.* Most voicers and wraps will end with an identification of the announcer, station, and location. This *sign-off* usually includes the reporter's organization and should help emphasize how close the reporter is to the story.

ON-LOCATION INTERVIEWS

When you are out on location at a game, event, or press conference, don't be timid. Ask for the interview and work to get it. The worst that can happen is that they can say "no." At best, you get a good interview and possibly an exclusive. Know about the person you are going to interview and about the topic or reason for the interview. If time allows, prepare a list of questions. Pay attention and follow up any answers that need more coverage. Get into the flow or rhythm of the interview and the answers will come easier from the person being interviewed. You only get one shot at the interview, so make sure your equipment is working and you are prepared. You may want to leave a pause before and after the answers and questions to make editing easier.

When doing on-location reporting, control your emotions. It is easy to get caught up in the excitement. A reporter needs to be an objective observer. Stay in control, but use some of the excitement in your voice and delivery. Don't make any unsubstantiated guesses, or report rumors unless they are an essential part of the story, and then only when they are identified as such. Identify yourself at the beginning of the report or know how you are going to be introduced. Know what hand-off will be used to get out of the report. Be prepared to discuss the situation with the anchor. If you are putting someone on the air, identify him clearly before you begin.

TELEPHONE INTERVIEWS

Telephone interviews are usually more direct than interviews done face to face. You need to ask questions that will interest the person so she'll keep the conversation going. Some radio and television stations may not have a telephone interface that allows the announcer to do the interview over the phone while working directly through the audio board. If the studio is not set up correctly for telephone interviews, here are some ways to insure a quality sounding interview. Make sure you have a good connection. If not, tell the person you are interviewing that you want her to sound good so you are going to hang up and try again. If you re-dial and don't have a better line, try to get help from the operator. When you start the interview, make sure you have the grip you want on the phone to avoid movement and the sound that goes with it. Talk into the telephone and also into the microphone at the same time. This will give your voice the studio quality and let the person on the phone hear your comments and questions. When you aren't speaking on the phone, move the mouthpiece away from your mouth to avoid breathing noises. Carefully articulate your words and avoid any slurred sounds. Slow down the pace and make sure the person you are interviewing knows what pace will sound best. Sound confident and don't interrupt the other person, especially with meaningless comments like "uh huh," "hum," or "OK." Work the audio-board levels carefully to keep consistent levels. If it is possible to equalize the telephone line, it should be used to give the audience the best possible sound.

TELEVISION INTERVIEWS

When tracking a possible sports story for television, the reporter should work to get the details of the story and any visual possibilities. Television is a visual medium and if the visual shots aren't there, the story may not work. The reporter should plan each part of the story, thinking out each element and how it could look. It is important to get the job done as quickly as possible so the material can be returned to the studio for editing. The quicker the story is done, the fewer last-minute headaches for the producer. The reporter should know how much time is available for the story and if it will be a self-contained package or if an anchor voice-over will be used.

Camera Shots and Editing Cues

The television sports reporter may end up on-camera in a variety of situations. A *cutaway* is when the reporter is shown listening or pretending to take notes during an interview. The camera typically moves to the opposite side of the location to frame the shot from the subject's point of view. Care must be taken to get the reverse angle correct or the shots won't match up. If the subject is looking right, the reporter must be shown looking to the left. Both the reporter and the camera person must pay close attention to the visual continuity. The cutaway shot allows the editor to place or edit remarks internally by covering any jump cuts with the reporter's video, making the transitions smoother. A *reverse question* is essentially the same question asked during the actual interview, but with the camera now in the reverse angle to view the reporter. This shot shows the involvement of the reporter with the story. While it isn't technically necessary for the subject interviewed to be there, ethically they should be present to make sure there were no basic changes in the question. A visual lead-in to a story is called a *set-up shot*. It is a view that includes the reporter and the subject in conversation. It can be shot before or after the interview is completed. It is used to set up the sound bite (audio) of the actual interview itself. The anchor or reporter can use this shot as a visual transition for the narration before getting into the actual interview.

In the field, a reporter must send back the fullest possible instructions, so the editor or producer can select the appropriate materials. Indicate all the lead-ins and out-cues and where they are located on the tape. Log the footage numbers or use the SMPTE standardized time code if available. Always count down into each segment to allow ease in cuing while letting the video recorder stabilize and avoid any distortion or tracking problems. Time will always be a major factor for a field reporter, so use in-camera editing whenever possible. Try to shoot the story in the appropriate order to minimize the time needed to hunt for the proper materials.

On-Camera Stand-Ups

When a field reporter appears on-camera, it's called doing a *stand-up*. There are three basic versions of stand-up. A stand-up "open" is at the beginning of a story to establish the reporter's presence and set the scene for the story material to follow. It is used extensively on live electronic news gathering (ENG) stories. A stand-up "close" is more commonly used than an open and is used to review important information given during the report or to add some new information to the story. A stand-up "bridge" is when the reporter appears in the middle of a story to help the viewer understand what is going on, or to tie together various elements of the story. The bridge can clarify and dramatize the story, if not overused.

The purpose of television sports news is to show events and information in more detail than the viewers would get if they were at the scene themselves. It is important to get people involved in the pictures and the story. People relate to people, not things. To convey the meaning of an event takes more than showing the event itself. People's faces are the best visual conveyers of meaning. Getting a reaction shot of people often means the difference between a flat report or one to which the audience can relate. The pictures by themselves should be able to show the story.

NEWS OR PRESS CONFERENCES

It is not unusual for a sports reporter to cover a news or press conference. Coverage of a news conference is only as productive as you make it. If the subject of the news conference is known, do some preconference investigation. If the text is given out, read it carefully. Look for any facts or story angles. See if there are any inconsistencies. Work out a list of questions or areas you feel should be covered. Make sure the questions are worth asking. Some reporters ask redundant or low-information-value questions just to show the boss they're working. Get to the conference site early and get a good position. You have a better chance of asking a question or following up and answer if you sit near the speaker. Note the time the conference begins and ends. During the presentation, look for any nuggets or items of information with which you can build a story. Check your watch each time there is a significant remark or question and make a notation so it will be easier to locate the material after the conference. If the tape recorder footage counter is visible, note down the number for later reference.

When you cover a press conference, make sure your equipment fits the situation. Your equipment must be battery-powered to avoid fighting for electrical outlet space that may not be available. Have all your equipment and cords clearly labeled to avoid confusion and lessen the possibility of theft. For radio coverage, you will need a cassette recorder with fresh batteries, a clean labeled tape, a microphone with a stand, and a long microphone cable. There may only be room to get your microphone on the table or speaker's podium with your recorder on the floor. Take along some silver duct tape and tape down your microphone, stand, and cords in the best possible positions. A foam pad under the microphone stand can keep extra noise and vibrations from interfering with your sound quality. If television stations are covering the press conference, see if you can put your station call letters on the microphone, but they may give you a hard time and remove the letters. Try taping the letters around your microphone so they cannot be taken off.

If you are doing TV coverage of a press conference, the lighting is very important. A well-run show will have adequate lighting. If not, all the stations may pool their lights or every person may work independently. Use a tripod to keep your camera still and level during the conference. There won't be room to move about. Get there early enough to put out all cables, microphones, power sources, and lights and test them. Make sure the camera station logo is visible so the person in the press conference can look into the lens when asked a question by that station's reporter. The question should be jotted down and a cutaway shot (picture of the reporter asking the question) taped after the completion of the show. The cutaway is edited into the report to give the interviewer extra credibility and visual impact.

PLAN YOUR VISUAL APPEAL

When you are at the scene, shoot any material that could be of interest to the story as the audio track can be laid down at a future time. The audio and video must supplement each other and work together to build the impact of the story. Try to plan the order of shots you want and then shoot them in the order you expect to use them, if it's practical. Write down a list of what shots you have taken and the cues to begin and end each section of material. This will make it easier to edit when you get back to the studio. Work to get the most out of the video you shoot. Use action shots when possible and look carefully for a background that will add information and visual excitement for the viewer. If you are interviewing a person, the shot and angle you chose are very important. Each camera angle and type of shot (such as a facial close-up or head and shoulders) conveys a certain meaning to the audience. If the person being interviewed is mad or upset, a close-up of his face might best pass along the emotional content to the audience.

After the videotape has been recorded, the reporter will go back to the studio and edit the material into the desired order and add any audio needed to complete the story. Build the story in your mind, based on the visual and audio materials you have gathered. In most cases, a reporter and a videographer will go out on a story. The two must have a clear understanding of what the intent of the story is to be. A good videographer-and-reporter team will use their combined knowledge and expertise to achieve their desired story . If the reporter is to do a stand-up (report, open or close on-camera), the picture and materials must fit with the rest of the story line and materials. The story from beginning to end must pull together to leave the audience with a solid image and understanding.

When taking action shots with videotape, film, or a still camera for slides, the videographer or photographer must plan what shots are appropriate and will add to the story. Pictures taken from up on the side of the court or field will give the best overall view of the game. If the angle is from the sideline or endline at the court or field level, the camera person must anticipate where the action will take place and plan accordingly. Let the action come to you and run into your field of view and then capture it. Most photographers work to capture the emotion and action of the sporting event. "Up close and personal" is very expressive and gives the pictures power and impact. Slowed-down motion (as shown so well by NFL Films) gives the games a certain grace and artistic effect. A different shutter speed or lighting change can add dramatic effects that will appeal to the audience and enhance their enjoyment and attention. Every story has an angle or reason for being on the air. The pictures should emphasize and build on the angle or story line to gain maximum impact on the audience.

Each television station has a news/sports video library or file. This material can be used when a story suddenly pops up or there is no practical way to get new and visually exciting materials. If a player announces his/her retirement, action footage of the player can gain the audience attention and boost their understanding of the story. TV is a visual medium; the more pictures that work, the better. The station needs an easily accessible file system that will cover all important sporting events, people, and related situations. Money deals, drugs, injuries, and strikes have moved into the sports arena and must be dealt with. Time is always an important factor in getting materials ready for a broadcast. Everything must be easy to locate and to prepare for use. Graphics and any artwork should be prepared to emphasize the story visually in order to have a stronger impact on the audience.

Each story needs a catchy beginning, logical buildup of the story line, and a memorable ending. Time will always be a factor in any story. You must know how long it should be and when the audience will see it. You may be working under a deadline to have the material ready to go by airtime. Foresight, planning, and concentration are needed to get the job done right and done on time.

3

Scoring with Information

ESTABLISHED INFORMATION SOURCES

One of the keys to being a successful sportscaster is the ability to find and utilize the information you want and need. A basic source of information comes from the wire services, Associated Press (AP), and United Press International (UPI). The wire services offer a wide variety of scores, stories, features, and polls. Both services have an audio network that offers shows, reports, and scores. AP also has a slide service for television stations. If you want and can afford another service, there is the Sports Network and the SportsTicker. Both services provide, among other things, scores of games in progress within minutes or even seconds of the sports happening. They are designed to supplement the AP and UPI sports wires. Another source is Computer Sports World, which offers a computerized on-line database of sports information. They have over 3,300 files of historic and current sports information with scoring updates as they happen. They will set up a customized file based on a station's needs.

Various audio networks offer a wide range of game information, reports, features, actualities, voicers, commentaries, and actual game play-by-play. There are regional and statewide services and networks to provide more selective local or regional information. The three major television networks, ESPN, CNN, and other networks provide game coverage and information. The major networks send video packages of sports down the network line for use by their local affiliates. Permission must be obtained to use game footage or selected information and pictures to avoid any copyright problems if you are not an affiliate of the source network.

A lot of written material exists that can prove to be very useful. Press guides are published by many teams and schools. Game programs are a valuable source of information. Be sure to ask the sponsors in the program if they would like to further sponsor their team on your station. There are newsletters, newspapers, plus specialty and general sports magazines. Teams have media days where the players and coaches meet with the media representatives to provide interviews and information. Some teams or schools will have an informational telephone line to provide reports, interviews, and other related information.

INNOVATIVE SOURCES

Aside from the established sources of information, the best sources may be the ones you discover and develop. It may take time and effort to locate and establish a working relationship with a good source. At the high school level, you work with the coaches, athletic director, and principal. Know who the leading supporters and boosters of the team are at all levels. Locate former coaches and players who can provide background and depth. At the college level, there is the sports information director (SID) whose job is to take care of the media relations for the team. A good SID is valuable to both the school and the media. The SID's boss is the athletic director who runs the whole show. Good relations with the coaches and players remain important. Professional sports have media relations people who are the basic contact between the team and the media. They help set up interviews and answer questions. On the amateur level of sports, you first have to find out who is in charge and willing to talk. Amateur sports coverage is important but can be time-consuming and frustrating, due to a lack of organization.

If it is possible, don't become totally dependent on just one source of information. If she can't or won't help you, you may get left out in the cold. You also need a second source to confirm information you have received. This is vital to avoid false information and a possible lawsuit. Make sure the facts are right or you may face big trouble. Locate and work with the area coaches, athletes, sports historians, sports-equipment-store personnel, sports medical personnel, and any experts or authorities in the various sporting fields. Build a file of names, numbers, and expertise; use their information and expertise when a story comes along. The more people you know, the better the chances are of finding the information you want. You can put a local angle of interest on most stories with the proper connections. It feels good to beat the competition and come in with an exclusive or just a solid piece of journalism. Your boss and the audience will appreciate it. Be tapped into the rumor mills in your area. Find out what the gossip is, even if you don't use any of the information. In many cases, where there's smoke, there's fire.

Try to establish a working relationship with your competition. Know all the media personnel in radio, television, and newspapers in your area and in the areas of interest to your listeners or viewers. Find out what the competition has discovered to avoid being left out. A close second with a story is better than no story at all. You may be able to work a different story out of the material your competition has used. In many smaller markets, the radio and newspaper people work together on stats and other information during a game. A friendly rivalry is fine, but if both can benefit from the association, all the better.

The sources of information are as varied and plentiful as you want them to be. Work out a system that will give you the flexibility and control you need to efficiently use the information you gather. Remember, sources can be burned very easily. You may hurt someone's feelings or pride and not even realize it. If you are given confidential information, respect it. You may have to say things about a performance or coaching job that the person won't like. Nobody likes to be criticized, especially in public. Keep your actions on a professional level. Part of your job is to cover both the good and the bad. You have to tread the thin line between putting a story on the air and losing your access to and help from the person you may be criticizing. Try to cover the ground rules before any confrontations or misunderstandings occur. Educate the coaches and athletes as to your job and any possible conflicts. Make sure they have a way to respond to your criticism and present their side. Give them an escape valve for frustrations and pressures created by their jobs and possibly by your actions.

There are many groups and organizations that can be of great assistance to a sportscaster. On the sports scene, write for information and get on the mailing lists of sports organizations and governing bodies such as the NCAA, AAU, U.S. Olympic Committee, NJCAA, and all league and sport organization offices. Make sure you know your state's high school athletic organization and how it functions. The more mailing lists you are on, the more information will come your way.

For the sportscaster, most states or regions have their own sportscasting and/or sportswriters' association. Many of these groups can be helpful in solving common problems, plus the extra clout of an organization can be of more use than operating as an individual. On the national scene, there is the American Sportscasters Association. They maintain an active job bank for sportscasters, have established a Sporsscaster's Hall of Fame, and publish a newsletter called *Insider's Sportsletter*.

4

Writing for Sports

ACCURACY

One of the most important areas of any story is accuracy. One of United Press International's mottos is "Get it first, but get it right." This is excellent advice for all journalists. Check and then recheck all facts, figures, and names. The time to make corrections is before a story goes on the air. The writer has a commitment to tell the whole story. A broadcast writer can't tell a 1000-word print story in six lines, so it becomes necessary to summarize. Don't let extraneous details get in the way of a clearly told story. Get the facts right, but don't overload the story with them. If the story is complex, it is better to summarize clearly than add elements that cannot be expanded or enough background given because of the medium's time limitations.

FORMAT

Use a standard size 8 ½ x 11-inch sheet of paper that is soft and will not rattle when handled or moved. Type on one side of the page only. Type only one story to a page. Scripts for television are normally typed in "books" of paper interleaved with carbon sheets. Each page may be a different color and distributed to the appropriate person by color code (brown to the anchor, pink to the director). Typing margins should allow for a 65- to 70-space line for radio and 35- to 40-space line for television. These margins make timing a sportscast easier. With a 65- to 70-space line, the average reader covers about 15 lines per minute (about 4 seconds per line). The television margins will allow the announcer to read about 30 lines per minute (about two seconds per line). Television scripts are typed on the right half of the paper, which allows for video instructions to be inserted on the left margin. Most television stations also use video prompters for the sportscast, necessitating that the script be typed on about half a page. Video prompting is a method of scanning a script and projecting it on a one-way mirror system mounted on the studio television camera. This allows the announcer to appear to be looking at the camera lens while reading the script.

All copy should be double- or triple-spaced. Use a normal capitalization style with upper- and lowercase letters because it is easier to read. Skip lines rather than indent for a paragraph so that it doesn't throw your timing off. Never split a word between two lines. Never split a sentence between two pages. Avoid starting two successive lines with the same word. If a story runs on to a second page, start looking for a place to split it after the tenth line. Centered near the bottom of the first page, type "more," circle it, and draw an arrow to the right edge of the page. To indicate the end of a story, triple-space and type the # (pound) sign or the word "end" with a dash on both sides in the center of the page. Do not number the pages until just before airtime in case the story order is changed. Number the stories, not the pages. If there is a second page of a story, it would be page 6, page 6A, and so on. Use paper clips to hold the script together. Don't staple the pages until the script is to be filed away. Sports stories should be dated and slugged with the writer's name placed in the upper right- or left-hand corner. The *slug* is the information at the top of the page that gives the name of the story, the program for which it was written, the date, and the name of the writer. These instructions may vary from station to station, depending on their individual requirements.

EDITING

If the copy needs to be edited, do not use copyediting symbols. Use a pencil to completely mark out any words that are to be deleted or corrected. Type or print the correction above the material crossed out. Retype any copy that is difficult to read. Circle anything not meant to be read aloud. Underline words that need to be emphasized. Mark a wavy line under any words that may cause difficulty for the reader, usually for meaning or pronunciation.

Never write a quotation into a script unless it is absolutely essential to the story. Paraphrase every statement you can. Quotes that must be read by an announcer may confuse the listener. If you must us a quotation, do not rely on the quotation marks in the copy to clarify things for the audience. Even when the reader is skillful in using inflection in her voice to indicate the start or end of the quote, the audience can still be confused. Although the announcer began a quotation with the word "quote," ending with "unquote" is not necessary. The inflection of the announcer's voice helps the listeners to understand that the quote is finished. If at all possible, paraphrase the statement unless there is something distinctive about the word or words to be quoted. If you are quoting someone, make it short. Avoid quoting someone with four or five sentences. If you use a long quotation, the listener might not understand when the person being quoted is done speaking and the sportscaster begins to speak his or her own words.

STYLE CHOICES

Write times in the regular way: "3 o'clock" or "3:25." Don't put "10 past" or "a quarter till." Minimize the use of the word "today." The assumption is that the broadcast news and sports is today's news and sports. Do not use "A.M." or "P.M." Use "this morning," "last night," or a similar expression. Use present and present perfect tenses as much as possible to reduce the need to specify times. For times more than one hour, don't use terms such as "90 minutes;" use "one and one-half hours." Don't use the phrases "This is just in" or "We have just learned," unless the information is less than 5 minutes old. Don't describe an event as happening "just a few minutes ago" unless it happened less than one half hour previously.

Skip all courtesy titles except where they are needed to avoid confusion, such as a story involving a man and woman with the same last name. Most titles of individuals and names of organizations should be stated in full the first time they are used. In subsequent references, you can shorten the name or title. Very long titles may be shortened. Avoid combinations and don't make place names into adjectives in titles. Be particularly careful not to use such adjectives at the start of a story.

Avoid using numbers as much as possible. When numbers are needed, simplify them and round them off. Try to analogize sums to things the audience is familiar with. For example, "300-feet long" could be "as long as a football field." Keep numbers out of the leads to stories and out of the first parts of sentences when possible. If you must use a number as the first word of a sentence, write it out. For example: "Twenty-five men were on the roster" or "on the roster were 25 men." Always write out the numbers one and eleven. Except for the number eleven, use digits from 2 to 999. Combine written numbers and digits for all higher numbers. Write numbers between 1100 and 9900 as hundreds, such as "83-hundred-24." Write millions and billions as "8 million dollars" or "3 billion dollars." Write out all fractions such as ⅕ as "one-fifth." Write one-place decimals as tenths: "3 and three-tenths," not "3.3." Convert decimals to regular fraction equivalents if they have them ("three-fourths," not ".75"). Avoid decimals of more than one place. If they must be used, use digits and write out "point" like "9 point 32 pounds." Write out percentages as digits and write out the word "percent," ("82 percent"). Use "point" with one-place decimals as well as longer ones ("9 point 3 percent").

Omit addresses except when they are an essential part of the story. Group numbers in addresses in normal speech patterns by using hyphens: "19-0-2 North Main." If the street name is a number, separate it from the house or building number by a direction or some other written material. Write "111 West 134th Street" or "number 722 on 22nd Street." If it is awkward or impractical to separate the house number and street number, write out the shorter of the two: "10-25 Tenth Avenue." Always write out First Street and Eleventh Street because "1st" and "11th" may be confusing to read. Omit ages unless essential to the story. Write all dates as ordinals: "December 15th, 1987." Some writers break the year with a hyphen to indicate the normal way of speaking. Write out dollars and cents for all sums of money. Write "3 dollars and 25 cents," not "$3.25." Round off all sums when possible unless the exact amount is vital to the story.

Use "a" or "an" instead of "per" in expressions such as "miles an hour." Do not use "a" in expressions such as "a hundred" where it might be confused with "eight." Always write out "one hundred." The only abbreviations that should be used are Mr., Mrs., Ms., and Dr. Write everything else out. Some abbreviations, such as Ft. for Fort (Ft. Wayne) or St. for Saint (St. Louis), may be used for commonly known place names. Very well-known organizations may be referred to by their initials. If each letter is to be pronounced, use hyphens, not periods, to separate the initials, for example: N-F-L. Well-known acronyms should be written without hyphens.

Use contractions only when they are easy for the reader to say and there is no chance of misunderstanding. Avoid the use of "n't" contractions except for the few that differ clearly from their positive form such as "won't" and "will." Negatives must always be stressed. Underline or separate negatives by three punctuation marks so they will be set apart. The three punctuation marks serve the function usually served by the comma. Each indicates a pause in reading. Do *not* use semicolons (;), exclamation points, (!) or brackets ([ ]). The comma indicates a short pause. Writers sometimes omit commas where no pause is desired in the reading. The dash (—) indicates a more complete break in thought, a longer pause in reading than with the comma. Be sure to understand the difference between a dash and a hyphen. The hyphen is a single line with no space on either side. The dash is two lines, preceded and followed by a space. Don't use a dash to hold together a string of words that should be a complete sentence. The ellipsis, which is three periods (. . .), indicates an even longer and more complete break than a dash. Use hyphens to make combination words easier to read, such as multi-talented or co-operative.

PRONUNCIATIONS

Always provide a pronunciation guide for any word that has any chance of being confusing to the reader. Underline the word with a wavy line. Write the guide in parentheses after the word like this: "telecast (tel-i-kast)" or print it above the word. Do not use diacritical markings such as those used in a dictionary. Indicate accented syllables with uppercase letters. Wire copy has no lowercase letters (unless newsroom computers are used), so the accent is shown with an apostrophe at the end of the accented syllable (TEL´-I-KAST). Others will use capitals to show the accent (TEL-i-kast).

The wire services provide pronunciation guides for most difficult words in their stories. They also run a daily list of names, places, and words in the news with a phonetic guide for pronouncing them. A good up-to-date dictionary provides correct pronunciations of most of the words you will encounter. Avoid guessing at a pronunciation. If you have made every effort to find a pronunciation and have failed, make one up that seems likely to be correct. That way there will at least be consistency in the pronunciation used until you can find the correct one. Avoid using words like *bow*, *lead*, and *read* which can be pronounced in more than one way, depending upon the meaning. Find out the local way of pronouncing local names and places to avoid sounding foolish and from out of town.

If you fail to find a way to pronounce a word, this guide from the Associated Press will cover most contingencies.

AH-like the **a** in arm
A-like the **a** in apple
EH-like the **ai** in air
AY-like the **a** in ace
E-like the **e** in bed
EE-like the **ee** in feel
I-like the **i** in tin
Y-like the **i** in time
OH-like the **o** in go
OO-like the **oo** in pool
UH-like the **u** in puff
KH-a guttural sound
ZH-like the **g** in rouge
J-like the **g** in George

Vowels are pronounced alike in all languages except English and German. Not all foreign names are pronounced as they look. Some words will defy any logical phonetic pronunciation. In such cases, use the rhyming technique. Thus, "James Bought (like hot)" shows how to use the rhyming technique. Don't expect to be letter-perfect in the pronunciation of foreign words. If possible, save the announcer some work and try to find ways to write the story without the tough-to-pronounce words, or at least limit them as much as possible. Remember that many foreign words, especially place names, have been anglicized, such as Paris is "*Payr*-iss," not "Pah-*ree*." Give thought to whether your audience is likely to have seen the words you are using. You may need to make some reference to unusually spelled words to keep the audience from being confused. Find out what local pronunciations are used to fit in with the audience.

CONTENT STRUCTURE

Broadcast sports leads (first sentences) should be structured in such a way as to make the listener want to pay closer attention. The lead is the broadcast writer's headline that enables the listener to get ready for the information to follow. Broadcast news leads generally emphasize what has happened or what is happening. Broadcast news leads seldom include names. The listener must be set up to pay attention for the name, unless it is widely recognized. Avoid asking a question in the broadcast sports lead. The broadcast sports copy should answer questions. Look for the "what" of the story. This will give the writer the attention-getting information needed. The time and place are occasionally included in the lead as long as the emphasis is placed on the "what." Avoid phrases or clauses in the lead that may add to the complexity of the sentence.

Don't lead a story with an unfamiliar name. Drop middle initials, except when the person is known by the name with the initials included. Initials may also be used to distinguish between individuals. Put any nicknames in parentheses. This will indicate to the reader that he can use the nickname or ignore it. Use the nickname only if it is very well known or unique. Use pronouns sparingly. Repeat nouns to avoid ambiguity. Don't use words that create ambiguity like "former" or "latter." Other words to watch out for are strings of sibilants such as "Commissioner Strauss's suggestions." Avoid alliteration, the repetition of words beginning with similar vowels or consonants. Watch for words that sound alike, but have different meanings. Check word combinations for the same problem.

The sportscaster may have to write a lead-in to a report that is on audio- or videotape. It may be used to go to a reporter or directly to the interview with the news source. The lead-in is used as a setup into the story and has many of the same characteristics as the straight lead. With the lead-in, the writer must be careful not to "steal" the story. Too much information should be avoided. The lead-in should identify the voices or faces contained in the audio or video report. Keep the lead to 10 seconds or less. Too much information in the lead may cause information to be repeated in the story. Avoid using similar statements in the lead-in and the beginning of the report to avoid "parroting." The tag or lead-out is the method used to wrap up a story and provide separation between individual news items. The tag can also reestablish the identity of the person in the taped interview or provide additional information not found in the story.

A television writer must fit the words with the pictures. Care must be taken to avoid overwriting or underwriting for the visual. Don't describe what the viewer can see. Set up the viewer for the scene and then wrap up the story at the conclusion of the visual. If the visual story is covered by a reporter on the scene who has already set up the scene for the viewer, a tag is not necessary. There may be times when a tag by the on-camera announcer is needed to separate the stories and keep a smooth transition between items. Do not write transitions as part of the story. This makes it too difficult to rearrange stories if needed. Arrange the scripts in the order they are to be used on the air, then write transitions on separate sheets and insert them in the proper places. Brief transitions may be penciled in at the top of the page of a story, provided the material is easy to read. Make each story in the script lead logically to the next one. This will minimize the need for transitions. A well-arranged script may need no transitions at all.

Since sports news is targeted to a specific audience, there is less need to worry about background information. The words used should be clear enough for all sports fans to understand. A more liberal use of slang and undefined terms is usually permitted. One of the most important areas is always the scores. Many stations pick specific scores or any unusual performances of interest to attract the hometown fans. Some stations will list only the winners while other stations will run through the whole list of scores so the listener can pick out any scores of interest and ignore the rest.

Sportswriters have devised countless variations of saying that one team has defeated another. After team one beats team two, many writers resort to: *edged, crushed, dumped, slammed, downed, clobbered, bombed, whipped, annihilated, embarrassed, wiped out,* and so on. Check with a thesaurus to obtain more words that can be used to describe games for as much variety as possible. Action verbs are nice, but stay away from the more sadistic or exotic ones. Even in the heat of a close race, it can sound silly. Another problem area is that of the cliché, the trite phrase or expression. Baseball has spawned an extreme amount of clichés. "It's a game of inches" or "one swing could send the game into extra innings" are two examples. This type of approach should be used very sparingly. It's easy to adopt the "sports language" approach and talk in terms of the "hardcourt," the "gridiron," with balls becoming "apples," "roundballs," or "bullets." Balls are seldom hit, but frequently "socked," "belted," "ripped," "smashed," or "pounded." Sports writers should have the goal of using clear, precise, meaningful, and picturesque language that emphasizes the action of the game.

5

Laws, Rules, and Sportscasting

Every sportscaster (and broadcaster) should know the laws and rules that govern what is allowed to go over the air and where trouble areas may occur for the individual and the station. Think before you speak. Put your brain in gear before you open your mouth. Once you've said it, you can't take it back. When a statement could injure a person or group's reputation or image, be very careful. Avoid any slipshod, indifferent, or careless reporting. The facts of a story should be confirmed and verified as far as practical, using usual procedures. Truth is a defense, but good intentions in reporting an untruth is not. There is a vast difference between what is true and what can be proven to be true, especially to a jury. Make reports of arrests, investigations, and other judicial or legislative actions precisely accurate, full, fair, and impartial. Limit any comment or criticism to matters of the public interest based on the facts that are fully stated and are true. Try to get the other side of any story. Don't color the story with your enthusiasm or opinion. Avoid any gossip and any unauthorized use of names and pictures for advertising or promotional purposes. If an error has been made, always have legal counsel handle demands for retractions, which come from a lawyer for a potential plaintiff. A well-meaning but unnecessary or poorly worded correction may actually prejudice a defense in a subsequent lawsuit.

DEFENSES AND DAMAGES

There are several privileges or defenses in any defamation action. The oldest axiom states that truth is a defense to any civil action for libel or slander. A privileged statement is one that, but for the occasion on which or the circumstances under which it is made, would be defamatory and actionable. The privilege concept evolved as public policy to further the right of free speech in a matter of public or social interest. In the *New York Times Co. v. Sullivan* case (1964), the Supreme Court decided that the 1st and 14th Amendments guaranteeing freedom of speech and press prohibited a public official from recovering damages for a defamatory falsehood concerning his official conduct unless he could provide that the statement was made with actual malice, meaning there was knowledge that the statement was false or there was reckless disregard as to whether it was false or not (Bruce W. Sanford, *Synopsis of the Law of Libel and the Right of Privacy* [Cleveland: Baker, Hostetler and Patterson, 1977]).

This idea was further extended from public officials to public figures. It is possible to be a public figure by injecting oneself or allowing oneself to be drawn into a public controversy, thereby becoming a public figure for at least a limited range of issues. The most litigated question in sports-related defamation is whether or not the plaintiff is a public figure. If so, actual malice on the part of the defendant must be proven before a recovery would be allowed. In order to show actual malice, the evidence given by the plaintiff must be clear and convincing. When the standards of investigation and reporting ordinarily adhered to by responsible organizations are met, damages may not be awarded.

Are coaches and teachers in high schools and state-supported colleges public figures or private figures? The courts have reasoned that public employees are hired and paid by public funds and are in highly responsible positions in the community in which education is a prime governmental responsibility. Public school systems, their athletic programs, and those who run them are consistently subjects of intense public interest and substantial publicity. Therefore, the conduct and the policies of public school coaches and officials are of as much concern to the community as are other public officials or figures. Since they are public figures (there may be specific exemptions so check with your lawyer), they must prove actual malice to win a libel or slander suit.

FAIRNESS ISSUES

There have been cases that demonstrate the privilege of fair comment in sports. When a jockey complained about a newspaper article that reported he had a cowardly and particularly unskillful ride in a race, the court noted that people who present their work or products to the public for its approval and acceptance are subject to public criticism and honest comment. In another case, a newspaper article stated that a coach who shouldn't be coaching anything was coaching a professional football team's receivers. The coach sued and the court decided that as a published opinion, it was privileged. The court called it a reasonable comment or criticism concerning a public matter and published for general information.

The court has offered some guidance on the subject of proper reporting. As long as the press correctly quotes another's statement about a matter of legitimate public concern, and does not distort it in any way and properly identifies the source, it may claim protection of the *New York Times* rule. It is also allowable to use hyperbole (extravagant exaggeration) and not lose the protection of the *New York Times* rule. You don't lose the privilege because it is allegedly "flippant" or "smart alecky" in style. It is wise to keep your remarks centered on the happenings on the court or field during a game play-by-play coverage. If something occurs that is part of the game action, you should run into few problems, aside from possibly rubbing some people the wrong way. It's always a judgement call on your part. In case of doubt, don't say it. Some announcers, especially sports talk-show hosts, have insurance policies to cover such situations. Move cautiously in controversial areas.

Be careful using any words that express or impute the following: a loathsome disease; a crime; words falsely charging arrest, indictment, confession, or conviction of a crime; any religious, racial, or ethnic intolerance; association with criminals; financial embarrassment; lying; membership in an organization that may be of ill repute at a given time. If you are threatened with a libel suit, or if someone demands a retraction, take the following steps. Receive all threats and demands politely and without any comment. If someone gets insistent, say the matter is out of your hands. Gather all the facts and contact management immediately. Don't admit any fault or error on your own or anybody's part. Don't defend what has been done. It's normal to be apologetic or defensive when someone is angry, but this could make matters worse. Don't release any copy, audio-, or videotape to anyone. Don't discuss what happened with anyone but management or counsel.

KNOW THE RULES AND PLAY BY THEM

The areas of litigation are constantly being tested and changed. New rulings occur that may change the interpretation of various laws. Don't take any chances; keep up with current rulings and judgments. In case of doubt, check with your management and station lawyer.

It is important to avoid any ownership or copyright problems. If the materials you want to use originate from another station or network that you are not affiliated with, you must obtain permission (written, in order to have proof positive) to avoid any copyright problems. Prior to the season or event, write and get permission from the station and/or network that controls the rights to the game or event. Once permission is obtained and the material you want is recorded, proper credit must be given to the source. In radio you give the source and in television a visual graphic of the source is displayed on the screen with the copyright material. If you are going to carry games or events from anyone, make sure you have permission in writing.

Know all the Federal Communications Commission (FCC) rules and regulations, especially those concerning an extended broadcast. Put a legal station identification within 5 minutes of the top of each hour, even though a continuous performance does not need to be interrupted for a station break. If you are running a delayed broadcast, begin with a disclaimer such as "this game was prerecorded for broadcast at this time." Establish your copyright to the game broadcast even though it may not be needed. With the current rulings on obscene material, care must be taken to avoid any language or phrases that could be interpreted as obscene. The new guidelines have established areas as obscene that are different from the basic seven dirty words that were the traditional check. These areas have yet to be fully defined and tested in the courts. Check for any rules that might affect your broadcast. An example of a rule that is from another source is from state associations that don't allow the promotion of any alcoholic beverages during the broadcast of a high school sporting event.

There has been a tremendous growth in women's sports due to Title IX, plus changes in social and personal values. This increase in women's sports activity brings increased broadcast coverage. With the majority of sportscasters being male, special care needs to be taken to avoid any sexist remarks. Sexual remarks or innuendos of a sexual nature should be avoided. Learn to use neutral remarks and expressions when referring to members of either sex. At the same time, don't get carried away to the extreme. In women's basketball, is it a man-to man defense or a one-on-one defense? The traditional man-to-man version is usually acceptable, but each announcer must make his own choice. Try not to offend the audience or the participants with your choice of remarks and comments.

Sexual remarks can lead to personal embarrassment, criticism, and a loss of professional credibility. Your station may be subject to harassment, sponsor pressure, and even legal entanglement. This reaction can have a very detrimental effect on a broadcast career. Comments about drill teams, cheerleaders, band members, other participants, or audience members involved in the broadcast should be carefully considered. Women sportscasters have the same responsibility. If you make an inappropriate comment or remark, consider whether to apologize for it or just go on as if nothing had happened to avoid calling attention to the words. There is nothing wrong about commenting on a pretty girl or a handsome man, or showing pictures of them on a television broadcast. It may be how you make the remarks that can cause a problem.

6

Sportscast Performance

It is common to dislike our sound or reflection because we neither see nor hear ourselves as other do. Most people don't believe that they sound like the voice that comes back from the audio recorder. The reason is simple. We hear ourselves speak through both the air and our bone conduction. The sound waves that come from our mouths are what others hear. The individual hears the physical vibrations that go through the bones of the head to the tympanic apparatus of the ear combined with the air-conducted sounds. Only when we hear ourselves through air conduction alone do we truly hear ourselves as others hear us. We are used to seeing ourselves head-on as in a mirror. TV shows us how we look to others, but we are not accustomed to these views, so we tend to like them less. Television can also distort our appearance to some extent. Most of us tend to look heavier on TV than in real life. Physical characteristics seem more noticeable. Remember that the audio and video recordings are only a surprise to ourselves. Others accept the way we look and sound, just like they accept us in person.

OVERCOMING FRIGHT

Because of the fear most people have when approaching a microphone or camera, adrenaline is released into the bloodstream, which causes one to become more alert and more energetic. If used correctly, the microphone or camera fright can be an asset. A person who is keyed up tends to generate more positive energy. Excessive nervousness can seriously impair a performance. Physical tension, shallow breathing, dry mouth, constricted throat, rapid heartbeat, upset stomach, shaking hands and knees, in any combination, may be present. These conditions may cause your voice to go up in pitch, or you to run out of breath in the middle of a sentence, have your voice break, lose concentration, read or speak at excessive rates of speed, or adopt a subdued attitude. If it becomes too extreme, you may not be able to communicate at all.

Before you go on-camera or talk into the microphone, remove yourself (physically if possible, mentally if not) from the scene. Take a deep breath, relax, and gather your thoughts. Narrow your concentration to the job at hand. Do a quick mental rehearsal if possible. Keep your message or objectives clearly in mind. Assume you are speaking to one or two people whom you respect and with whom you want to communicate. Talk to one person, even if there is a crowd or you are on-camera. If you look at the camera or the microphone, assume it is the person with whom you are carrying on a conversation. If it is a camera, look into the lens, not above it or off to the side. You are establishing eye contact with the audience (the one person you are talking to).

Concentrate

Concentration is vital for any performer on radio or television. Like listening on the telephone with a poor connection, you must concentrate to separate the information you want from the surrounding noise. You must learn to give a topic your undivided attention; any small distraction might destroy your point of focus. Use your mental rehearsal and practice. Concentrate on different plays and work on how you should describe them. The time to feel the part and generate the emotional content in your voice is during the practice and rehearsals. The ability to summon emotion depends, to a large extent, on the development of your imagination. You must be able to imagine a world of appearances as a real world and hear and feel sensations from past experiences as if they were in the present. If you can develop such an imagination, you can create emotion in yourself and transmit it to others.

It may be difficult to keep your attention where it belongs during a game. There are many things going on that can distract you. You are the eyes of the audience and need to have selective tunnel vision. The audience needs a continuous flow of information to stay in the game and enjoy it. There will be times when you must split your attention among several things. You may have to watch a play, make a mental or physical note, describe the action, and plan for the next movement. Split attention and concentration are essential for TV play-by-play when calling the game action and listening and responding to the director's calls. Keep in mind the needs of the audience. It gets easier as you practice concentrating on the essential. Once you have established good habits, you will almost automatically zero in on the important items needed by your listening audience.

Detach

It is important for the professional announcer to create a second "you." You must learn to detach yourself mentally. Learn to stand back mentally, look at yourself critically, analytically, and dispassionately. You can become your own producer and director. You can learn to reserve part of your mind for this function. Your mind will act like a puppet master, controlling and guiding your performance. This helps prevent slipshod or uncoordinated work. If you make a mistake, thinking about it can lead to further mistakes and spoil your delivery. If you make a mistake, your physical voice continues to read while your mental voice says, "That was silly, but now you know what happened so it won't occur again." Then you dismiss it from your mind and concentrate on what you are doing. Don't let one mistake, which the audience has probably already forgotten, ruin the rest of the performance.

Mentally Rehearse

Mental rehearsal is not just casual daydreaming. Learn to practice in your imagination or mind's eye how to describe various actions. Practice games and plays in your mind and build a set of mental reflexes that cover the basic action situations. Take a specific play or action and work it over and over again in your mind. Use different combinations of words and phrases until you feel comfortable approaching a specific play or action. Work on the combinations and specific situations until it is second nature to you, an automatic reflex. Research has shown that imagined events imprinted deeply in the mind are recorded by the brain and central nervous system as memories. With enough repetition and practice, one can imagine well enough that the memories are not distinguishable from actual physical experience. Practice things in your mind and build a set of mental and physical reflexes. With enough practice, you can go into a "cruise control" mental state during a game or event. This frees up part of your mind to concentrate on other aspects of the game broadcast. With practice, you can improve your mental reflexes.

Believe in Yourself

You must believe in yourself and your abilities. Many things happen that can shake your confidence. Technical troubles, mistakes, and personal criticism can create frustration and damage your self-confidence. No matter what happens, don't give up. Whether you feel confident or not, you must sound that way to the audience or lose credibility. You are the expert eyes and ears of the listener. You must be believable. No matter what happens, give it your best shot and then move on to whatever is next. The more experience you have, the easier it will be to shake off mistakes and problems. You are in the same situation the athlete has to face. When he makes a mistake, he doesn't stop but keeps on giving it his best effort. If you want to be a winner, you must do the same.

AD-LIBBING

Successful ad-libbing is one of the hardest areas to learn. There are many times that you have to fill with material when there is little or no game action. Plan ahead as far as possible. Write down key words that you can use as a cue and reminder. The more points of interest you can put down on paper, the easier it can be to focus your attention from point to point. Try to put them into the order of desired use. Make sure the material is relevant to the situation. Give the audience what they want and need to hear.

BREATHING

A good announcer or reader must know how to breathe correctly. Many people use their upper chest muscles. This leads to shallow breathing and is hard to control. Learn to breathe with your diaphragm, the muscles just below the rib cage. Lie down on a solid surface, such as the floor or a bed. Place a fairly heavy book on your stomach. Practice breathing so the book will move up and down as you breath. Practice this form of breathing and it will become a habit and a natural way to breath. Diaphragmatic breathing gives you the control and the power to move air at the rate and force you desire. You control the movement of air over your vocal cords and mouth with more precision.

Breath through your mouth and not your nose. The rush of air through your nose creates a noise and the smaller openings don't allow the volume of air to move that breathing through your mouth allows. Before you start reading or talking on the air, take a medium-size breath, not a giant breath. Breathe in short controlled bursts of air, with short intakes of air at logical points. Try not to go so long that you sound strained and are forced to take another big breath. If you are reading material, place diagonal (/) marks between the words or phrases that are logical breathing points in the copy. These visual cues will help you plan your breathing and make your reading sound more natural and less forced.

READ NATURALLY

Reading aloud is vital to anyone who is planning to be on the air in radio or television. A good announcer must know how to read materials in a natural sounding way without mistakes. Always review and practice the material before you deliver it to the audience. If you don't have time to review the material and must read it "cold," slow down your reading pace, and work your way through it carefully. Always work your eyes ahead of where you are vocalizing. Your eyes should be three, four, or more words ahead of where your mouth is. This method will allow you to discover any mistakes or problems with the material and give you a chance to adjust and make corrections without pausing or reading the mistakes on the air. This also allows you time to prepare your mouth and breathing to fit the needed words and phrases, allowing a smoother transition between words and phrases.

Reading materials on the air must be natural sounding. A good reader seems to be thinking and uttering the ideas for the first time and not merely repeating words. One of the first steps in expressing the meaning is to understand the meaning of what you are reading. You must master not only each literal and explicit meaning as in the dictionary, but also the implications or implied meanings. People normally do not speak in single words, but in groups of words. The meaning of a line is based on the proper grouping of those words. Word groups are used when expressing a single idea. A typical word group usually contains from six or eight to a dozen words that are tied together to express a single idea.

When conveying ideas, you must be capable of centering or focusing attention on some part of the thought unit at the expense of other, less important parts. You need to set your mind to putting the exact shade of meaning into the whole sentence or word group. The stressing of individual words will usually take care of itself. The reader must be responsive to any attitudes or emotional values that are built in or inherent to the lines. The emotional quality of the words will color the meaning as much as the logical content of the message. The reader must be responsive to the meanings of the words as they are verbalized. The material must be read with the ring of personal conviction and belief to have any credibility.

One trick actors use to add more flavor or color to the lines is to *think into* the words additional words or phrases that express the appropriate implications. For example: "I'm satisfied [no more than I was the last time];" or "I'm satisfied [I've got to be, my cause is hopeless];" or "I'm satisfied [satisfied that you are a liar]." This list could go on and on. Add words that express the meaning of the words you are going to read out loud. You may wish to practice using the extra words to perfect the intended meaning.

There are several common faults that readers encounter. One problem is "running down hill" during the reading. The opening part is hit hard and them the end trails off. This sounds monotonous and also emphasizes the meanings in the beginning, but not at the end of the line. When a reader is under tension or is excited, she tenses up and speaks at an excessively high pitch level. She also tends to speed up the pace of her reading, leading to mistakes and possible audience confusion. The best speaking level, as to the quality with a maximum volume and a minimum effort, is usually slightly lower than that of the ordinary speaking tone. Most readers need to slow down, lower their voice pitch, and then overemphasize their mouth movements when reading. Exaggerate your facial features when you read aloud. This will build into a habit that allows your face, and the related sounds, to fit what you are reading aloud. Make sure you clearly utter each sound in the word and don't slide over sounds, especially near the end of the word.

VOICE CONTROL

Your voice is the game or action for the audience. The impression the audience has of you depends mainly on your voice qualities. It is important to know your voice strengths and weaknesses so you can practice and improve any areas of weakness and polish even more the areas of strength. Look into a mirror and pucker your lips while keeping as small an opening as possible and say, "How are you today? I'm fine." You will notice that you have changed the resonance or the amplification of your voice. Resonance is created by the vibration of the sounds of your voice through 18 different structures in your body from the base of the rib cage on up. The pucker test will help you find a resonance quality focused in the back of your mouth. There are other places to focus your voice for good resonance. You must learn to control the placement of sound in your oral cavity and consistently choose the location of your best sounding voice. One of the most common resonance problems in the United States is nasality. To find out if you're nasal, put your fingers on the bridge of your nose and say, "A, E, I, O, U." If you felt any vibration, you are probably a nasal speaker. All the English vowels are supposed to be produced in various parts of the mouth.

The tongue is the primary director of resonance. It has many ways of dividing the mouth cavity and causing sounds to take on many different qualities. A good speaker will learn to vary the resonators instead of just increasing or decreasing volume. Both chronically loud speakers and those who are not loud but maintain constant intensity have tense vocal mechanisms. The loudness or softness of a person's speech level is connected intimately with self-image. Often it is best to increase intensity without increasing loudness at all. An important step in managing your loudness level is to make sure you are willing to share the emotions that change in loudness and intensity.

Regional accents may contribute to an individualistic voice but can make the speaker sound uneducated, dull, or boring. The listener believes the inadequacy may be the person rather than the voice. Dullness for the average voice usually means not changing speech rate, loudness, intensity, or voice pitch. Most people prefer voices of a lower pitch compared to higher pitched voices. A rate that is too fast usually accompanies a higher voice. There are several techniques used to find your natural pitch. Many times it is just a bit lower than your normal speaking voice. One technique used to find your natural pitch is to take a deep breath exactly as you do when yawning and say aloud, "Ahhh . . . I know that." Do it again, this time making the "Ahh . . ." the most pathetic tone you can manage. Now do it one more time and this time listen closely. The sound "that" ends on will be your natural pitch. Once you have discovered your natural pitch, you must learn how to control it. Poke around in both a high and low voice until you feel you know where all the sounds are coming from. Once you know how the muscles feel at both the high and low ends of your range, you will begin to feel uncomfortable when you start to move out of your natural voice area.

When you hear a light, wispy, breathy voice, or raspiness, consider how the lips and jaw are moving. It is not uncommon to hear bad quality when the lips are barely moving, the jaw is not relaxed, or both are held in a strange or unusual position. Mouth movement is very important to the quality of your sound. Practice good movement of your lips, jaw, and facial muscles by overemphasizing everything when reading various sounds. Exaggerate your facial movement to an extreme while watching in a mirror. Notice where and how your face and jaw move with certain sounds. With practice, you will get your face into the practice of moving to emphasize different sounds naturally and smoothly. If you have a stone face, lips, and jaw, you will not have a good sound. Use a round object, such as a pen, and place it between your front teeth. Then read materials while holding the object between your teeth. Exaggerate your facial muscles and lips to clearly articulate the sounds. Repeat the exercise with a larger object, such as a cork. This forces your facial movements and lips to overreact to articulate the sounds.

Timing and quality, as well as pitch, will create any bias or changes desired by the announcer. The right level of loudness presents the speaker as a strong capable person. It enhances your natural pitch, requires a good breath support from your abdomen, and is resonated easily from your best point. Learn to vary the speed of your speech to take advantage of both short and long words. Jerkiness is a distracting timing characteristic. A jerk is a quick and suddenly arrested movement. It lacks coordination or rhythm, which is usually very noticeable. English is very rhythmic with a regular recurrence of certain aspects or beats. People are inherently rhythmic and they will immediately react when that rhythm is missing.

TELEVISION PERFORMANCE

The television performer's audience is a small intimate group that is sitting about 10 feet away from you. The camera sees everything you do or don't do. It will look at the pleasant and unpleasant details. You must learn to control your actions without letting the audience know you are doing so. Make sure you know the composition of the shot. Know which camera is for a close-up and which is a cover shot or longer view. The closer the shot, the less you should move to stay in the correct area or you might get clipped. Don't move toward the camera, let it come to you unless told otherwise.

Movements

Television announcers are usually shot in medium close-up or close-ups that do not allow room for large gestures. Avoid gestures out toward the camera; it will have the effect of exaggerating the size of your hands in relation to the rest of your body.

Reserve your hands for meaningful movements calculated to assist viewer understanding. It is common form for sportscasters not to use any form of hand gestures. Their hands are normally out of camera view or seen holding a script. Hand gestures would suggest emotional involvement beyond a point that might be acceptable for this type of broadcast. If you use your hands on TV, avoid insignificant gestures that will be a distraction to the viewer. Make any movements smooth, steady, and a little slower than in real life.

When standing on-camera, avoid rocking from side to side. Stand still. Weaving or rocking from one foot to the other can be distracting when on a long shot and disastrous on a close-up. In the studio, there is usually a monitor placed where you can see it; thus you know when the camera has a wide, medium, or close-up shot. You will be aware immediately if you are moving out of the picture. In the field, you may not have a monitor and know the composition of the shot. The best approach is a *minimum* of movement at all times, except when it is called for. To reduce the tendency to rock, stand with your feet slightly apart and with one foot turned out to form a 15- to 20-degree angle with the other foot. The turned foot should be several inches in front of the other foot. This position makes it very hard to rock.

While you are sitting on-camera, it is easier to avoid excessive random movement. Remember that most movements are exaggerated on television. Work to reduce upper torso or head movements that are rapid or wide-ranging without reducing your natural energy levels. Sideways movement can be very annoying, especially on close-ups. Movement toward and away from the camera can take you in and out of focus. Be sure to sit up straight and don't slump while keeping your head up.

When rising, sitting, or moving about the studio area, you must move somewhat slower than you ordinarily would. Begin a movement with a slow and slight motion, pause, and then follow through with the intended movement. Camera operators are trained to follow motion, but a little thoughtfulness can guarantee that you will stay in the shot. You should not sit down or stand up on-camera unless the movement has been planned in advance or instructed by the floor director. If the camera is on a "head shot" or a standing person and they suddenly sit down, the head drops out of the picture. When the camera is on a head shot and the seated performer suddenly stands, the viewer gets the infamous "crotch shot."

When sitting or standing near another person in an interview, you may be positioned closer to the person than you would if you were just carrying on a conversation off-camera. In American society, we are surrounded by an invisible area we consider our own psychological or personal space. Most of us sit or stand at a comfortable distance. Television is no respecter of psychological space. The intimacy of television is best shown when the interviewer and guest can, at times, both be seen in a medium shot. To sit or stand too far apart is to force the director to settle for a close-up or wide shot.

Cues

To have a show properly paced, accurate cues from the floor manager/director are needed. Different stations may use slightly different cuing signals or procedures. Review the cuing system with the floor manager before you go on the air. React to all cues immediately, even if the cue doesn't appear to be appropriate at the time. A professional performer will react to different cues quickly and smoothly. Don't nervously look for the cue; wait for the manager to draw your attention to the cue sign. Don't acknowledge the cue in any way. They'll know if you got the message or not. Visual cues from the floor director may include time cues, directional cues, or audio cues. Here is a small sample list of cues used by most television stations with only minor, if any, variation.

Cue	*Meaning*	*Hand Signal*
	Time Cues	
Stand by	Show about to start	Extend arm over head and point to camera
Cue	Show is on the air	Point to the performer or the live camera
On time	Go ahead as planned	Touch nose with fore-finger
	Directional Cues	
Closer	Performer must come toward camera	Hand motions waving toward the floor manager
Stop	Stop or stay there	Extend both hands in front, palms out
O.K.	Well done, continue	Form an "O" with fore-finger and thumb
	Audio Cues	
Speak up	Needs to talk louder	Cup hands behind ears or move right hand upwards, palm up
Keep talking	Don't stop talking until next cue	Extend a thumb and forefinger horizontally, move like bird beak

Facial Gestures

It is important to know the capabilities of your face in regard to emotions or moods you are trying to convey. Some people with a more rounded, fleshy face can hold a smile that will look genuine while other faces trying to maintain a smile would look strained and unfriendly. Avoid "stuck-on" expressions or poor imitations of other performers' facial gimmicks. If you are left on-camera for a few moments, simply turn your head toward a real or imaginary monitor and let your face register anticipation. When directly addressing the camera (which is the viewer), look straight into the lens of the camera. Focus your gaze about a foot behind the glass lens. In a studio situation, look at the camera with a red light, called a tally light.

Eyes are very important in television. They show your thoughts and emotions. A dull, unimaginative person has dull, listless eyes. But when your eyes are lit up with enthusiasm, they convey interest and involvement that invites the attention of the viewer. There are some purely mechanical ways to show emotion to the viewer. If you narrow your eyes and contract your brows, you can give the impression of deep concentration and intense thought. Open your eyes wide to indicate surprise or wonderment. If you pause during an announcement, turn you eyes deliberately sideways and either slightly downward or upward, away from the lens, and then back again until you look straight into the lens, and continue speaking. This can suggest the idea you have arrived spontaneously. It takes practice and timing, or you may look furtive or apprehensive. It is much better for your facial expression to arise spontaneously as a natural reaction to your mood and thoughts.

When searching for a thought or word, many people tend to raise their eyes toward the ceiling as they pause for inspiration. This is usually distracting and unflattering. If you have this habit, work to overcome it. Avoid flickering your eyes around. Look steadily into the camera or toward the interviewer. Avoid rapid or excessive blinking. Don't freeze on a smile while waiting for the director to "go to black" (take your camera shot off the air) or to another camera. It is usually best to continue small and natural movements while you wait. Look off to the side, as if to a monitor, if you are going to another source. Nonverbal communication, such as facial expressions or head, hand, and torso movements that are not overdone can add much to your communication ability.

Clothing Choices

Before you go on camera, plan you clothing carefully. If the television station uses a chroma-key matting, you should avoid any shade of the color used for the chroma-keying, usually blue or green. Chroma-keying is the process that allows a picture from one camera to be keyed or superimposed into a portion of the picture from another camera. If blue is the color of the chroma-key background, a blue shirt or blouse would have a second picture appearing on the area of the blue cloth to the audience. Avoid clothing that has small checks or narrow stripes. The TV camera cannot handle fine, high-contrast patterns; a wavy shimmering picture, called a *moiré effect* can occur. Avoid black-and-white clothing. Pastel shades are best for nearly all broadcast purposes and are complimentary to people of all skin shades. People with extremely dark faces should wear clothing somewhat darker than those with light skin tones. Avoid excessive contrast between your face and clothing. Avoid clothing the same shade and color as your skin. Jewelry can cause video problems if studio lights reflect directly into the camera lens and cause flaring. Makeup can be used to reduce skin shine, eliminate 5-o'clock shadow, improve skin color, and hide minor blemishes.

Microphones

Prior to the beginning of the show, a small microphone (called a lavaliere) will be clipped to the shirt, coat, or tie. It picks up the talent's voice when they speak in a clear, natural, but fairly loud voice. Prior to the show, the audio engineer will ask for a mike check so they can preset the audio levels and make sure everything is working. Don't shift around in your seat or move extensively, as the sound of clothing rustling or a chair squeaking will be picked up by the microphone. All mikes are "dead" (turned off) until the director gives the cue for studio audio. Wait until you receive the opening cue from the floor director or your voice will be clipped and not heard by the audience. Don't take your opening cue from the turning on of the tally light on the camera unless you are specifically told to do so. A boom mike may also be used, a microphone suspended from overhead by a boom. Avoid fast turns, movement up and down, and don't speak down toward the ground. If you are getting up, do it slowly so the boom operator can move the microphone out of the way and yet keep the proper distance. If a desk or stand microphone is used, keep your hands off and don't move it, or tamper with it when it has been turned on.

When using a hand microphone on TV, be sure there is enough cable for any moves you are going to make. Handle it as carefully as possible. If it is dropped accidentally, make sure it is checked immediately to be sure it will work. If you are in the studio, don't hold it too close to your mouth. Hold it approximately chest high for good audio quality. If you are outdoors or in a noisy surrounding, hold the microphone closer to your mouth. If possible, turn your back toward any noisy areas to reduce unwanted sounds. Get a firm, comfortable grip on the mike and don't change hands or fumble around with it once it is turned on.

When you are interviewing someone with a hand microphone, don't push it into his face. His volume will usually be lower than yours, so get the microphone a bit closer to his mouth to help make up the difference. If you are in a noisy situation, put the microphone closer to his mouth. Know the position of the camera and the shot composition so you can hold the microphone where the person's face will not be obstructed from view. If you need both hands for a demonstration or other use, hold the microphone under your arm with the live side pointing as close to your mouth as possible. If you are off-camera, hold it steady and in one fixed position to keep the volume level.

If the video that accompanies your narration does not come up as planned, keep on reading the script. You must, however, change any direct references to the pictures. Carefully watch your monitor. As long as the audience can see what is going on, you don't have to tell them. Be sure to tell them what they cannot see on the screen. Be sure to identify the scene and the people on the monitor. Don't hesitate to make references to the screen if it helps to clarify your identification. Avoid using color as a reference ("the team in the red uniforms"). Some people receive color telecasts in black and white. The use of dark and light will work for all sets.

Like a good advertisement, the sportscaster's open and close are very important. If you cannot hook the audience at the beginning, the rest may go by unnoticed. You must set out some bait, an attention-getter or tease, to attract interest for the rest of the material. When you are finished, it is vital to leave the audience with a favorable impression so they come back for more. Write out, practice, and polish the open and close you plan to use. If you have a toss-off to verbally cue or pass attention to another announcer, be prepared and smooth-sounding. A quality sound doesn't happen by accident.

One of the finest forms of flattery is imitation. Listen and watch other sportscasters and find out the phrases, words, and styles they use. Take those words and phrases and use them or change them to fit your style of announcing. Adapt what you hear to fit your situation and personality. If an announcer is making major money and sounds good as a sportscaster, study what she is doing and take anything you can use. Examine her style, delivery, voice qualities, vocabulary, phrasing, and anything else you can think of. If you copy a certain style of phrasing of another announcer, make sure it fits with your style and the station's approach, or it comes across wrong and sounds fake.

7

Radio Sports Talk

A growth area in sports and media is the radio sports talk show. Many AM radio stations are unable to compete with the music FM radio stations. A way to counter the music and still attract a desirable target audience is to specialize in news and talk. The sports-talk format has been around for many years and is gaining in popularity. A person wanting to enter this job field needs a solid background in sports. You must be versatile and adaptable to handle a variety of situations. The ability to ask tough questions in a way that gets a response is important. You must be able to point out significant facts and interpret them in such a way that the audience understands. The job requires enthusiasm and vitality plus the ability to locate and capitalize on controversy. It takes a variety of skills and knowledge to be a successful sports talk host.

COMMON FORMATS

A radio sports talk host must develop the skills to conduct interesting and informative interviews with studio guests and the ability to talk with a wide variety of strangers who call in on the telephone. Most sports talk hosts work a maximum of 4 hours on the air, which is about the amount of time a host can remain sharp, energetic and patient. The show producer assists the host in selecting and scheduling guests while handling the correspondence and taking care of all guests. Topics for discussion that will be interesting or controversial must be developed and readied for air. The program opening must stimulate the interest of the audience and help motivate them to telephone and offer their opinions.

Most shows will start with a music theme. The host then ad-libs from a prepared opening that includes segments, and topics, plus any guests who are scheduled to appear. The opening comments should generate listener interest. No interest means *no* telephone calls. Most radio stations have a policy of the host not talking more than a specific number of minutes without taking a call, even if it's a fascinating guest. The aim of most sports talk shows is to maximize listener involvement.

Having studio guests is not required but adds variety to the program. Guests, including the most famous, are seldom paid for their appearances. Most guests agree to appear because they see it as an opportunity to promote a cause or interest of their own. Take the time to explain to the guest the nature of your show, the kind of listeners you are attempting to reach, and any relevant station policies. Once the guest has been scheduled, appropriate promotional materials must be prepared and used. Most stations keep a log or record of which guests appeared and how well they performed so future appearances can be considered.

The commercial announcements that run during the sports talk show will be clustered together. Sports talk shows cannot tolerate constant interruptions. Specific times are usually designated for when the commercial clusters will be aired. The host and producer must keep track of the time and make sure the sponsors get their commercials on the air. A program log containing the list of commercials is kept and checked by the show producer or host.

TELEPHONE TALK

A sports talk host seldom uses the actual telephone to talk with the callers. A telephone *interface* unit allows the call to be run directly through the audio mixing board, allowing the host to listen in the headphones and talk into the microphone. It is more convenient and provides higher quality for the listening audience. Some hosts wear a headset microphone to keep their hands free. It is important to work the callers in "real-time," so they are not distracted by the sound of their voice on the radio. The caller should be told to turn down their radio before they can be allowed on the air to avoid feedback and disorientation of the caller delay system.

The telephone calls received by the sports talk host may cause some problems. Callers become angry when they call and cannot get through to voice their opinions. There are the long-winded, persistent callers, or those who try to get a commercial or partisan message on the air. There is the problem of objectionable language and the possibility of slander. If it's on the air, the station and operator are responsible for it. There are also the callers who keep calling back, over and over again. They enjoy it, but the listening audience may get bored.

There are several ways to keep unwanted callers off the air. The use of a *screener*, a person such as the show producer who takes the incoming calls and stops questionable or objectionable callers from getting through to the show host, may work. There are ways to delay the telephone calls from getting on the air. One way is to use a *digital delay*. This piece of equipment holdsup or delays the output of the conversation for a predetermined time period, up to several seconds. You simply throw a switch to cut off any conversation before it gets on the air. Another way is to use two reel-to-reel tape recorders. Record the telephone conversation on one reel machine and play it back on the second machine. The time it takes the tape to go from one machine to another is the amount of delay. If the host decides the conversation should not be allowed on the air, the output of the second tape machine is turned off before the audience can hear it.

Because of the possibility of obscene or slanderous materials reaching the air, the host generally controls the "cut" button to stop material in the delay. Many stations use a tape-recorder log to retrieve and reconstruct precisely what was said on the air in event of threatened legal action. The programs can be recorded on one audio track while a telephone company time check is recorded on the second audio track, providing an exact time of all on-air events.

The show's producer or screener can use a variety of systems to signal the on-air host about which calls can or should be put on the air. Most stations utilize computers with specially developed software to fit each individual situation. The computer display screen indicates the nature of the call so the host can pick and choose the ones to be put on the air. There is usually a timer to monitor the length of the call. Many stations and hosts make it policy that a call should not last longer than 2 or 3 minutes.

When the telephone lines are opened, the audience expects to be entertained and informed. A way to access information must be located for use by the host. Reference books, magazines, articles, record books, and other information sources must be available for the host to find information and keep the "expert" sound. Personal computers have programs available to provide the sports talk host with easy access to information. The easier the access, the better for the on-air host.

Preparing for an interview with a guest, in person or on the telephone, is very important. There are nuggets of information ready for the interviewer to dig out. Those nuggets are easier to locate if you are correctly equipped. There are two basic types of interviews: those with a person for his information and those with a person for his personality. A combination of the two is the most desirable. Depending on the "why" of the interview, hard questions may have to be asked, or it may be just an effort to bring out their personality. Keep mentally on your toes and work to bring out entertaining stories and information. Be versatile and adaptable to each situation.

A major challenge for the sports talk host is to motivate calls from new or infrequent callers. Repeat the telephone numbers often on the air and tell the listeners which lines are open and what the discussion is about. Do *not* beg for callers. Most shows are structured in units that can be added, deleted, or reassembled into customized program segments that will appeal to the target audience.

A sports talk host may use humor, pacing, and interesting phrasing to create a program that entertains and appeals to the target audience. Most sports talk hosts consider themselves entertainers, not journalists. There are a variety of styles and formats used to attract and hold the audience attention. Some hosts act as investigators, some will advocate certain ideas, and some will deliberately provoke the audience with outrageous statements and comments. Whatever style is used, it must fit the personality of the host and the needs of the audience. It appears that over time, the factual presentation of information with expert interviews tends to draw higher audience ratings than the screaming, angry conversation approach.

By carefully screening callers, a station can help control its programming. Telephone calls are the playlist of the sports talk show. The screener constantly manipulates the lineup of incoming calls, placing them in order of interest or impact. The screener can filter out the "regulars" who call the station too often or those who cannot make a coherent contribution to the show. Most stations prohibit the use of a caller's full name to avoid imposters. When the time comes and there are no callers, a file of clippings or articles can be used to provide information for a monologue by the host. An "expert" telephone list should be developed with people having specific expertise in areas that should interest the listening audience. They can be used to fill time and generate audience interest and telephone calls.

The typical caller does *not* represent the typical listener. Callers represent a small fraction of the audience and differ from the program listener. Sports talk hosts learn to keep the needs of the listening audience primary in everything they do. Most sports talk shows will construct each programming hour to appeal to specific target demographics. Free-form or open-line programs, where the callers set the agenda, should be severely limited. Conversation programs are effective within the conversation environment because they reach an audience that is tuned in to listen.

8 Radio Network Sports

Network sports on radio can be very important to a station and its sports image. A well-produced sound with the so-called network quality adds to the distinct sports sound of your station. Network sports can add credibility to your station. It provides a profitable service that is appreciated by the fans. It can be a hot selling item that brings in major dollars to the local station. Major team sports generate enthusiasm that rubs off on the station. There are many promotional activities that may be associated with the teams and their sponsors.

TYPES OF NETWORKS

There are several types of sports networks. There are the national radio networks, such as CBS, NBC, and Mutual. They offer game play-by-play coverage, features, commentaries, and more. There are, however, not that many events that attract a national audience such as the Superbowl, World Series, or the Olympics. The most common type of network sports coverage comes from the team/regional/state networks. Major colleges and professional teams organize their own networks to carry their product. There are also specialty networks that will carry a certain event or type of event, such as the Indianapolis 500 race or the Motor Racing network that covers major American auto races. With the exception of the major networks, provided you are an affiliate, you usually have to pay for the right to carry the game or event. It is possible, in some situations, to carry a network sports event or show if the primary affiliate in your market doesn't want to carry the event or show.

COSTS AND SYSTEMS

The cost of carrying network sporting events depends on the sport, the delivery system, the team, and the situation. The network originator has to cover the cost of the production, the rights fee to carry the event or games, and the distribution to the network stations. The local station has to pay for the right to carry the event or game and the distribution of the product with money, or network commercials, or a combination of both. The network will try to pass along as much of the cost as possible and yet keep the largest possible number of stations in the network. The bigger the audience, the more they can charge for the network commercials that all stations have to carry.

There are several different ways that the network signal is delivered to the local station for broadcast. It may be a telephone line or loop that connects you to the *flagship* station. A flagship station is the station that originates the signal for the entire network. A fairly common form of delivery is the dial-up telephone system that has the local radio station calling a telephone number that connects them to the audio signal from the game. This is about the cheapest way for the stations involved, but the sound may lack in quality. The use of satellite feeds to stations provides the best sound quality but may be too expensive for many operations. Both systems will be discussed further in Chapter 10. Another way to cut delivery costs is to use a station relay system. Several high-powered FM stations receive the game feed from a phone line or other source and broadcast the game over the air as usual. Then other stations pick up the on-air signal and rebroadcast it on their own stations. The quality is good (depending upon the distance and terrain between stations) and the use of the FM signal avoids the interference that can occur with an AM signal from thunderstorms or other electrical interference.

FORMATS

The networks provide format instructions, schedules, times of games, and related shows and materials. Promotional materials and a list of network sponsors and their order or position in the broadcast are included to avoid local stations running conflicting sponsors during the game or event. The network sponsors may also prompt local sponsors of the same product or service to join in and take advantage of the network advertising.

The network format sheets include the opening billboard information, which includes the name mentions and slogans for sponsors of the event. Time is usually specified when the local stations insert their sponsor's information. Broadcasts usually open with introductory remarks and the billboard information given by the game announcers.

Specific out-cues are listed so station engineers or board operators know when a commercial break should be started. Out-cues such as "We'll be right back after this time-out" or "This is the Indiana University Football Network" are common examples. The exact time of each break should be listed in the format information. The network engineers add several seconds before starting after a break to make sure those along the network get back from their break and won't clip the comments of the announcers. For a 60-second break, the network engineers usually allow for 62 or 63 seconds before cuing talent to begin.

Commercial schedules are used to avoid conflicting products or companies during the same broadcast in the same break. The type of ad, duration, source of the ad, and, if local, the number of the carted commercial assigned to a recorded message are included. By placing all commercial material in a specific order, fewer mistakes should occur. Backup material, such as written commercial copy based on prerecorded commercials, should be on hand for the announcer or engineer. You can't charge a client for material that did not get on the air.

9

Marketing and Selling Broadcast Sports

There are excellent reasons for carrying sports on radio and television. Radio sports has a highly selective reach in the upscale, upper-dollar male audience. It has an ongoing pressure that builds up over a period of time, producing long-term results. You can adjust to a specific audience and their needs and wants. There is flexibility in the approach to the needs of the audience. Sports produces lots of enthusiasm and advertisers can become a part of that enthusiasm. It is a natural for the involvement between the advertiser and the audience. There are many areas that are open for marketing and promotional activities for the station, the advertiser, and the team/event. In radio, sports is equal to morning drive-time in efficiency. Stations carry fewer total commercials and with no direct competition for the duration of the game, listeners get involved with the sports, the station, and the advertisers. Stations can attract a new and different audience that may not have sampled the regular station format. New listeners, or viewers, may be picked up or at least made aware of the station and its programming. A good sports package can boost an audience during the day part of the event and effect the total cume (total number of different viewers or listeners) of the station. This increase can help the station's total numbers (ratings listing total viewers/listeners) and increase the attractiveness of the station to advertisers.

PACKAGE SELLING

When it comes to selling sports play-by-play events, there are many approaches to making money. One of the keys in sports play-by-play packaging is to sell the entire season as a package, not one game at a time. Selling one game at a time is fine for supplemental sponsors, but the season package gives the station the financial foundation to support the entire sports effort. Most sports are sold to sponsors in a package that may include commercials in the game or event, name mentions in the open and close game announcements, and promotional ads to run during the season with sponsor name mentions. The promos that run during the season get the sponsors' names out to a variety of listeners and viewers, which means you are getting paid to promote your sports programming.

It is better to sell two 30-second commercials or three 20-second commercials instead of one 60-second commercial. You can sell to more sponsors, make more money, and have a wider variety of commercials with less repetition. The 60-second break, or time-out, is the most common break in sports. If a sponsor wants a 60-second add to run, charge them more for it. Special locations during a game or event, such as just before halftime or at the 2-minute warning, can be sold for extra dollars if the sponsors are interested.

A current trend in selling local sports is using a full year contract that covers all local sports or a combination of sports. A total price and monthly payment is established, based on the number of guaranteed and possible games and events. This format provides a solid base of sports income while the advertiser can spread the payments over a full year instead of paying just during the sports season.

In addition to the regular sponsors, it is possible to sell ads to sponsors for specific games. They may not get the promos or open and close name mentions, but it is more money to be made and another sponsor to be satisfied. Make sure the regular sponsors know in advance that this may happen. It is also possible to sell *drop-in ads* that run during the game itself. Use written 10 to 15 seconds of copy for the announcers to read during the breaks in the action of the game or event. There are times when there are breaks in the action that are too short to break for a commercial, yet long enough to use fill materials such as the written commercials. Again, make sure the regular sponsors know what is going on so they don't get upset. Be especially careful with competitors or conflicting products. If you use a policy of exclusivity, have the price high enough to cover what could possibly be made from additional ads.

Have a halftime special available to sell to a sponsor or sponsors. You may want to do interviews, or send the program back to the station for the latest news and weather. Make sure you have a pregame and postgame show or wrap-up available for your sponsors. These shows offer extra availabilities for sponsors and add to the depth of your station's sports coverage and image. If more than one game is going on at the same time, and you are not doing play-by-play on a delayed basis, offer in-game reports or updates. Have a person at the other game to call in reports while the game is being broadcast. Have them take along a radio and listen for announcer requests to update the game, or go on the air after every commercial break in your game. This can be very effective during tournament time and adds greatly to the depth of your sports coverage. These reports can, of course, be sold to another sponsor.

Make sure the package you sell to the sponsors covers the games you plan to run on a *tape-delay basis*. If two games are on at the same time or overlap, record one game to be played back after the first game is completed. The audience interest is not lowered when you do this. If they were at the game, they want to hear how the announcers saw the action. If they were not there, they don't know what happened and it's as if they are at the game as it occurs. Two games in a row are fine, but more than that may be too much for your audience. You must know the situation and be flexible enough to please the audience and the sponsors. It is also possible to delay a game into a low commercial-rate time if there is a good enough reason. Many television stations take this approach to local sports and videotape games to be shown at a later time to avoid bumping a higher paying show or event. Daytime AM stations have used this technique for years. If it is warranted, cover pep rallies and other special events. These can usually be sold over the phone on a onetime basis. The station should jump on the bandwagon and become part of the scene and reap the benefits from the situation. Enjoy the great public relations opportunities these events offer your station.

In addition to the actual play-by-play coverage of a game, supplemental shows should be planned and sold. A pregame coach's show sets the scene for a game, plus it's a nice ego stroke for the coach. This show should be prerecorded if possible. That takes the pressure off the coach and the announcer before the game when both already have enough to do. It also allows the show to be edited and polished so it sounds or looks better. You can make a nice package of the interview, the open and close, plus commercials. It is a good lead-in to the game and can be sold for top dollar. Arrange for interview times (and phone numbers for radio) to avoid any logistical problems on either side. If something happens, you can still do the interview prior to the game and record it, or run it live. Make sure the coach understands the format and what is expected. This gives you a fallback position to cover any problems and keep the sponsor and audience happy.

SPONSORED NEWSCASTS AND WRAP-UPS

At halftime, the announcers need a break to get ready for the second half and finish their work from the first half of the game. One inexpensive way to cover the halftime period is to have a newscast from the studio. A sponsored newscast can be sold as a separate show that may appeal to a news sponsor rather than a sports sponsor. The news may be offered in a package of an open and close, name mention, plus one or two commercials inside the newscast. The news may be prerecorded or done live from the studio. This keeps the studio personnel involved with the game and allows the latest news, weather, and sports to be offered to the audience. Separate promo-tional announcements may be offered as an extra incentive and allow the price to be increased.

A postgame show or wrap-up sounds good and offers more sponsor availabilities. Make arrangements ahead of time if you plan on having the coach come to you after the game. If the coach is not available, have a wrap-up show with information about the game and possibly use audio or video cuts of the big moments of the game. It takes planning and work but sounds or looks great when you pull it off. Always record a wrap-up of the game from the scene immediately after finishing your on-air work. This allows the game announcer a chance to talk about the game while it is still fresh in her mind. The crowd noise, or just the ambience of the location, adds to the authenticity of the report. This also avoids forcing the announcers to rush back to the studio to do a wrap-up, when it can be tiring enough just doing the game. The wrap-up is designed for scoreboard shows and sportscasts. Depending on the use of the wrap-up, the length of the report must be closely monitored. More than one version may be recorded to give some variety to the scoreboard or sportscast.

Weekend scoreboard shows are very effective in attracting a listening audience and sponsor interest. They also enhance the sports image of the station. A well-produced scoreboard show will attract more than just the traditional sports enthusiasts. Use distinctive music and promote the show and its sponsor(s) just like the other shows. Make sure the studio announcer who is doing the scoreboard show knows which scores to read and what information to include in the show.

Obtaining Rights

When a radio or television station is interested in obtaining the rights to cover a major sports event or team, a careful investigation must take place. The station must determine if it is advantageous to carry a team or event, and whether the competitive advantage over other stations in the market is worth the effort and risk involved. The strength of a station's bargaining position depends on the motives and needs of the team or organization. It takes mutual cooperation to have a successful and financially profitable relationship.

PROMOTIONAL OPPORTUNITIES

There are two basic types of promotions, *topical* and *image*. Topical promotions push specific events or information, such as when the next game will be broadcast. The image promotions work on long-term loyalty and the emotional feeling the audience has for the station and its sports efforts. Many promotional messages use both the style, or theme, of the image promotion with the information of the topical promotion. Try to hit the audience on both the emotional and intellectual levels.

Sports on radio and television offer a wide variety of promotional and marketing opportunities. The level of the sport and the interest that is generated determines what might be done. High school players may be limited by their coach or school as to any activity related to the media. College players may be available, depending on their specific situation. Professional players are expected, usually for a fee, to be available for promotional efforts. If rights fees are being paid, part of the package should specify that appearances and participation in promotional efforts must be made by coaches and players. One point to remember is that amateur athletes cannot receive money or goods worth more than $5, except for trophies and plaques, or they could lose their amateur status.

Sponsors should be involved whenever possible. They should receive "point of purchase" support such as posters, signs, or stands. It may be possible for coaches and players to make appearances. Investigate having a photo session or team visit made to the sponsors. Tickets to the games should be distributed to the sponsors. The more they are involved, the better for all concerned. A contest for sports sponsors can help motivate action and generate enthusiasm. Sponsors can be involved with contests for the listening or viewing audience. An example is when fans pick up entry forms at the sponsor's place of business to enter a "Home Run Inning" contest. A name is drawn from the entries. During a specified inning, if a home run is hit, the contestant wins a big prize, or cash. If a triple, double, or single is hit, a lesser, appropriate prize is given. At the end of the season, all contestants in the game are entered in a drawing for another prize. This keeps up interest for the entire season.

Special events can be created in conjunction with or separately from the sporting event. Sponsor a 10-kilometer race that finishes under the goalpost on the football field prior to the game, with the participants receiving tickets to the game and acknowledgment by the crowd at the game. Extra activities can open the door for a variety of sponsors who like a certain type of activity. The more people you involve in a contest or event, the better. Adult pep bands or marching bands, bringing back former cheerleaders, any activities that fit the event and involve people should be checked and used. Keep looking for activities that sponsors, the station, and the target audience will enjoy.

Stations can run a tailgate party in the parking lot before the game for the fans. Sponsors may provide food or drinks or other materials for the enjoyment of the fans. Throw a barbecue or chili cookout featuring the coaches and players before or after the game. Have an autograph session with materials provided by the station and team. A photo day, posing fans with their team heroes, is great public relations for everyone. Run a contest or have a drawing to allow certain fans to shoot baskets with the team, hit in batting practice against a star pitcher, skate with the hockey team, or catch a pass from a quarterback.

Stations should use their own merchandise and novelties to give to the audience during the game, both to those who are present and those who are listening or viewing. Many schools run a half-court shot for money, or raffle off the game ball to raise money. It doesn't hurt to have materials for the schools to give away for your station. Good-looking banners should be shown at the game so the fans know you are there. Some stations use printed banners so they can be taken by the fans after the game and used as posters. This is a good idea for high school and college students. Bumper stickers should be used to promote the station and team. They should also be used in contests, possibly for game tickets or other incentives. An excellent source of promotional ideas can be found through membership in the Broadcast Promotion and Marketing Executives (BPME).

Have coaches, players, cheerleaders, drill team, and band members make appearances at station-sponsored events. Make your own drill team or cheerleader competition. Consider off-season events, such as football players against station personnel in a volleyball game for publicity and fun. Combine the coaches and players with the station "all-stars" to compete with local groups for charity. Off-season events can be created, such as the "battle of the network stars" competition. Keep your station actively involved and generating interest in the off-season.

Another way to make money and add to the station's sports image at a small cost is to establish a sports scoreboard/information telephone number using a telephone answering machine. Put out a local sports-line service of scores, game information, and other sports materials that will interest your audience and sell a promotional message to a sponsor and also plug your own station. The announcers at the station can update the information on a regular basis. Of course, any games you cover will receive top billing. It's one more way to promote your product. The calls are "free" to the public. Push the number during the game and during the regular air shifts. This service not only makes you money but can be used as a resource tool in sales and promotions. It can provide indication of the sports interest and impact you are generating in the area.

Create your own local team polls and standings. Newspapers have used this approach for years. Weekly picks and point spreads may be used to generate interest. Some radio-TV stations combine with the local newspaper to get additional coverage for all concerned while increasing their credibility. Pick an "all-area" or "all-star" team for each season and sport. Run a "player of the game" or "player of the week" award. Create news about the teams and players. Many stations and newspapers sponsor award dinners or give trophies at the end of the season. It adds to the station's image and leaves a good impression. These awards may cost a little or a lot but generate interest and a positive feeling. The sportscasters should attend all award dinners and personally present the awards for the station.

Use coaches, players, or cheerleaders to cut station liners and promotional messages for use by the station. Let them endorse your station and the coverage of the team. Get them to work for you! See about sponsor endorsements or mentions from the players and coaches. Some athletic organizations have special causes to promote, such as the NFL and United Way. Jump on the bandwagon and turn the effort into a positive appearance for the station and the benefit of the cause.

Many radio and television stations conduct coaches shows or interview shows in public places with audience participation and even telephone call-in questions and comments. Work out a deal with a local restaurant to provide space, food, and drink. Some restaurants offer a special deal for those who want to show up for the show. It generates business for the sponsor and lets the team's image rub off on the sponsor's name. Some places have sports event parties that can be sponsored or promoted by the station, such as Superbowl parties.

Get involved with the local boosters club. Arrange for airtime and station support for their fund raising efforts. The clubs have current and potential advertisers involved. Also check with alumni and other special interest groups, such as band parents. Work with the boosters club on contests and other events. They have the people if you have the idea and the publicity. It's good public relations to work with an influential group for the benefit of all concerned.

Some stations will operate a "fan bus" to take team supporters to away games. Build a package around the bus ride, such as tickets, seat cushions, drinks, cups, and so on. Target your activities for specific audiences, such as women. Special give-aways and contests can be designed to build interest and support. The fan bus is a good visible way to show support and keep everyone happy. Have station personnel go along to help control and entertain the fans on the ride.

Sports are packed with action and excitement. Promote the "thrill of victory and the agony of defeat" to your audience. On-air promotional efforts play a major role in the image and impact your sports coverage has on the audience. Play on the action and emotion that is naturally present. Make your promotions dramatic and exciting. Use music that fits your sports package and the overall station image. Build a common theme or basic style for other sports programming. Television stations need a visual style or approach for their sports efforts. A common theme and style gives the audience a consistent look that will be associated with the station's sports coverage.

On-air promotional efforts only reach current viewers or listeners. Other media, such as print or billboards, are needed to gain the attention of people who aren't normally in the audience. Every promotional effort should reflect the action and emotion of your sports coverage in the style or approach of your station. Try to dominate any medium you decide to use. If you advertise in *TV Guide*, spend enough to dominate the issue. Your efforts are watered down if you don't stand out from the competition.

Make the audience associate *your* station with *their* team. Make the team efforts rub off on your station. Become *the* sports station in your area. If you cover a winning team, jump on the bandwagon and ride along. People like winners and want to be associated with them. If your team is having a poor season (someone has to), find positive points, such as outstanding players, and build from that perspective. Have coaches, players, and others work for your station by cutting liners and providing interviews. Actualities from interviews can be mixed with music, play-by-play, and other production materials to create promotional messages. Have the cheerleaders do station cheers. Design a "cheer" contest for your station with local cheer teams.

On-air promotional messages, with and without sponsor mentions, must be run on a regular basis. Schedule promos to run with commercials during the game. Sell "your" product while you have an available audience. It shows the sponsors you value the time enough to use it yourself. Push other station sports efforts. Have announcers read promotional materials during the game in addition to produced materials. On-screen television graphics should be used, with or without announcer audio support.

Professional and college teams usually have a media day for reporters and station representatives. Take written lines for the players and coaches to read. Use their name value and glamour to enhance your station's image. If you carry network sports, a promotional package is usually part of the deal. They provide promotional ads, sportscaster's message, and music beds. Use this material and build promotional packages that enhance your part in the broadcast coverage. Many stations pay for customized promos to be recorded for their sponsors and their own station's sports efforts.

10

Radio Sportscasting Equipment

The equipment that is needed for a successful remote broadcast will vary with the situation and the delivery system to be used. The equipment picked must provide the best quality sound for the money you are willing to spend. There are some equipment items that are basic and essential to any remote broadcast.

Any station that carries sports must be concerned about the sound quality and how the audience perceives it. Just because sports do not originate from the station's studios doesn't mean that it must sound muddy, distorted, or just plain bad. A poor-sounding product is a turnoff for both the listening audience and the advertiser. It also reflects on the total station sound and image. In too many cases, stations try to "cheap" their way through the remote broadcasts and it sounds that way. The quality sound that sports should have must start with a management that is willing to support the sports effort with time, money, and leadership.

Sports are entertaining, exciting, and must be presented that way to the audience. The game package must fit into the station's format and image. Quality production must be done on all in-studio work, such as the promos, game open and closes, plus the commercials. The higher quality the product, the better the audience will accept it. When the announcers go out on the scene of the game or event, they need all the support they can get from the station. The open and close should have exciting, distinct music that will attract attention and help build an image in the mind of the listener. It will help enhance the audience's memory of the event. There are technical ways to help enhance the sound and those will be examined later in this chapter.

MIXERS AND MICROPHONES

One of the most basic pieces of equipment is the mixer/amplifier. It must be portable, durable, cost effective, and able to run without plug-in electrical power. There are many styles, types, and makes available with features and prices to fit most situations and budgets. The equipment's basic job is to mix the incoming signals from the microphones and other sound sources, and then amplify the signal and transmit it to the main studio for broadcast to the audience. This one piece of equipment can be very costly, but basically controls the sound that the audience hears.

Microphones are needed that provide a good quality sound and can stand up to loud noises and possibly bad conditions. A headset microphone is comfortable, convenient, and leaves your hands free to handle other chores. There are different styles and types available for you to choose from so you can get the best possible sound that fits your situation and budget. A microphone on a stand is common but can be bumped and knocked over. It is inconveniently placed in front of you where you may need the space for something else, such as your scorebook or other written materials. Always keep an extra handheld microphone hooked up and ready to go for interviews, crowd noises, music, or as a backup in case anything should happen to your other equipment. Use the extra microphone as one more safety feature. Bring spare connectors and microphone plugs to the game/event and the tools to make any repairs that could save your broadcast.

Types of Microphones

Microphones change sound vibrations (your voice) into electrical energy. The electrical energy goes into the recorder, mixer, or other electronic device. There are three basic types of microphones used professionally: moving coil, ribbon, and condensor or capacitor. The moving coil microphone (also called a dynamic mike) is the most commonly used type of microphone. It has a good sound quality, is durable, and can stand loud noises and fairly rough handling. A ribbon microphone produces the best quality sound but is rather delicate and cannot take loud noises. The capacitor microphone reproduces a brief crisp sound very well but needs a power source, such as a battery, to work. All three types are used in studio work, but on a remote broadcast, the moving coil is the most commonly used.

There are several special-purpose microphone applications used in connection with sportscasting. A headset microphone is a moving coil microphone (with a built-in "pop" filter to help remove the hard "p" sound) that is attached to a pair of headphones. The microphone is usually unidirectional (picks up sound from one direction) toward the mouth to keep down unwanted noise. This style of microphone and earphones keeps the announcer's hands free and allows movement of the head while staying on-mike.

In television studio work, a small, unobtrusive microphone called a *lavaliere* is used. This microphone is small enough to clip on to clothing or hang around a person's neck. This microphone provides a quality sound and has a built-in high frequency boost that is needed since the chin cuts off many of the higher frequency sounds. Lavalieres are usually omnidirectional (pick up noise from all around) due to their location when in use.

There are two microphone styles that are used to pick up noise from a distance, such as a court or field site. The shotgun microphone cancels the sound in all directions except for a narrow angle at the front. This cuts out extra noise in the area and allows concentration on a desired sound source. The quality of the sound may be sacrificed for an apparently greater reach. The parabolic microphone consists of an omnidirectional microphone that faces inward to a parabolic reflector. Instead of cancelling unwanted noises, the parabolic dish (which is concave) concentrates the desired sound waves from the source and directs that sound into the microphone.

Another sports microphone that can be used is the wireless microphone. It consists of a microphone attached to a battery-powered transmitter that will feed the microphone signals to a receiver that is hooked up to the mixer/amplifier. The wireless is perfect for placement where microphone cords would be impractical, such as near the sidelines of a football game. The quality is usually good, but may be subject to interference from other transmission sources, such as CB or police radios.

Extra Microphones

Always hook up an extra microphone at the game site. Use that microphone (usually handheld style) as a backup in case the headset microphone should go bad during the game. The spare microphone can also be used to do any interviews, pick up the sound of the bands, the crowd, or the national anthem.

Each event has its own use for the extra microphone. In football, the microphone is needed to pick up the sound of the crowd and their reactions to the game. This gives the game sound greater excitement and credibility. The audience can enjoy the feeling of the game through the sounds created by the crowd and the play. A remote or wireless microphone is perfect for football coverage. A microphone at the field level can pick up the crack of the pads, the thump of the kicked ball, and efforts of the players. It also adds greatly to the quality of any halftime band shows that you are carrying. Quality remote, or wireless, microphones are well worth the money it takes to get the field-level sounds on the air.

Using the extra microphone for baseball is much like the use of the microphone in football. A microphone placed near the backstop will pick up the crack of the bat on the ball, the thump of the pitch into the catcher's mitt, and the sounds of the umpire and the crowd. The sounds of the game add greatly to the enjoyment of the listening audience. Be careful where you place the microphone so it won't be hit or pick up the wrong sounds. Keep the microphone away from any public address system speakers. Always have a long microphone cable or wireless microphone available, and know if you need a stand to place it on. Use silver duct tape to firmly anchor your microphone wherever you plan to use it. If it a windy day, place a foam windscreen over the microphone to cut down on the sounds of the wind whistling or moaning into the microphone. If you don't have a foam windscreen, use a sock or a mitten. Be very careful when mixing the sound from the extra microphones into the announcers' sound levels. It is very easy to drown the announcer's voice level if the crowd is yelling or even if the wind is blowing hard.

Basketball is different than baseball and football with respect to the noise and the use of microphones. You are located in a building and the crowd noise is usually too much, as opposed to too little. You put out an extra microphone, but only for use on an interview, or for the national anthem. The announcer's microphones will pick up plenty of noise created by the crowd. If you are at a game with a very small crowd, you may want to use the extra microphone to add some noise to the broadcast.

Audio Levels

The levels of the crowd and the game noise to that of the announcers' voices must be carefully controlled. The announcers' voices should be two-thirds to three-quarters of the total sound the audience will hear. Check with your engineer or audio board operator to make sure the levels are correct. It helps to listen to the on-air signal to adjust your levels properly. If the crowd noise gets too loud, or the pep bands get fired up, you can cup your hands around the microphone and your mouth to help cut out the extra volume of noise. The microphone level (how sensitive it is to noise) should be turned down; the announcer then puts more energy and volume into his voice to make up the difference. Be very careful not to over-drive or over-modulate the signal from the microphones to the mixer. Too much signal will sound distorted and mushy. Be extra careful placing extra microphones, due to loud or very reflective sound sources.

The lack of noise can be much worse than too much noise. If there is little or no crowd noise, especially in a gym or closed area, it will sound extra loud and seem to echo to you. Announcers tend to talk in a hushed, soft voice. It can seem embarrassing enough to make you lose your voice energy and enthusiasm. It's fine to drop the volume of your voice, but only if you keep the energy and enthusiasm. This style is also used when covering golf and tennis.

TV Microphones

In television, the number of microphones and their placement depends on the needs and resources of the telecast. Many stations will place a person on the sidelines with a directional or parabolic microphone to follow the play and get the sounds of the action. In basketball, the backboards may have microphones attached to them to pick up the sounds of the ball on the rim and board. Other microphones may be placed to bring the realism of the game to the audience. An audio engineer controls the total sound mix of the microphones in use. Both radio and television broadcasts can use their microphones to create a stereo effect. Some stations will put their announcers on a left or right channel and have the crowd in the middle, or have the game sounds work from right to left.

FIELD REPAIRS

Every sportscaster must understand the equipment and know how to make repairs in the field. Equipment problems will occur. A small tool kit may save your broadcast. It could include a knife or wire stripper, needle-nose pliers, small vise grips, both types of screwdrivers in different sizes, and any other tool you or your engineer think could be of use to you. A small soldering iron may also be included. One item that should always be taken is silver tape, also called *duct* or *gaffer's tape*. The tape can be used to tape cords and equipment down for safety and security. It also works as electrical tape for connections and wires. It can even be used to hang your banner and fix your radiator hose on the way home in your car. It is a useful item at a very reasonable cost. Have the station engineer give you a fix-it course and write down notes to take with you to the game on how to do the various repair jobs.

LISTENING TO YOUR GAME

If you are in range of your station signal, you need a radio and connectors to hear the game as it is broadcast over the air. If you hear what the audience is hearing, you can better judge the quality of the signal, the levels, and any problems that occur. It is always nice to know your game is on the air. The radio should be a powerful model to give you maximum hearing range. It should run on both plug-in electrical power and on batteries. It must have an output connector so you can plug your headset in and hear what is happening. If more than one person needs to hear the broadcast, a *splitter* is used. A splitter is a connector that takes a single output from the radio and splits the signal into two or more outputs for the headsets. A splitter with individual level controls is very handy since different people may want the output levels at different volumes.

If you are out of the receiving range of your station signal, you need to run your breaks by using a timing method. Have a standard or recognizable out-cue for the board operator so she knows when to start the commercial break. Use a stopwatch or regular watch and time the length of the break. If it is a 60-second commercial break, start talking at 62 seconds. This keeps the show sounding tight and professional. It is the board operator's job to start things on time and bring up the remote output level correctly. If you are on a system where you can talk to the board operator during the break, check in at each break and make sure you are still on the air, and find how the levels of the broadcast are being handled. Do not take it for granted that you are still on the air; always check. One very valuable tip to anyone who is using this technique (or any time you may be on the air) is to *never* say anything you don't want to go out over the air. Always assume the board operator missed the break and left your microphone on and going over the air to the listening audience. Avoid saying anything that would prove embarrassing and even possibly cost you your credibility or your job.

Phone Hookups

Depending on the remote delivery system to be used, different items will be required to make the system work correctly. If you are using a dial-up telephone system, you need a phone with a voice coupler that hooks the output of the mixer into the phone line. Some mixers come with built-in telephones. If a phone is already in place and doesn't have a voice coupler and you cannot replace it with a phone of your own (such as multiline telephones), you can unscrew the mouthpiece of the phone. Hook your mixer output wires to the two rings or prongs that are in the telephone mouthpiece. Have a connector wire prepared by putting two alligator clips on one end to hook to the telephone, and a dual plug or other connector that will fit with the mixer unit. Know how to fix and repair telephone equipment. Learn which wires are important. Usually the red and green wires are your best bet. Modular telephone hookups present a much bigger problem. It takes a special tool to put a modular plug on to the line. The loop and the RTU systems take a direct feed from the mixer.

Recording Equipment Setup

If the game is to be recorded at the scene, use the best recorder and tape you can find. Feed from the mixer to the auxiliary input on the recorder. Use a top quality tape of 90-minutes' duration (45 minutes per side). Make sure the tapes are cleanly erased and have usable labels on both sides. Clean the heads on the recorder before it is to be used. Use plug-in electrical power whenever possible. Make sure your batteries or power packs are fully charged and ready to go if the power fails or is not available for use. If your batteries run down during the recording, it will speed up the tape sound when played back at full speed and you will sound like Donald Duck. If the game is to be recorded at the main studio, use the best reel-to-reel tape available, not the junk reels of tape that are always around. Make sure everything is clearly labeled to avoid any confusion. Check and double-check everything to make sure it is working smoothly and correctly. It doesn't hurt to stay a bit paranoid on the subject of the equipment working correctly or not.

COMPUTERS

Computers are being used in game coverage situations. There is a set of programs called *Statman* that is designed to produce game and player statistics as the game is being played. It is available for amateur and professional baseball, basketball, and football. Cumulative team and player statistics are updated immediately and are easily accessible through the terminal. It is fast enough to allow the cumulative stats to become part of the play-by-play coverage. The program can be operated by one person inputting information provided by spotters. The operator can be trained in two or three practice games. The cost of the program depends on the sport. The program is compatible with most major personal computer brands.

MISCELLANEOUS EXTRAS

There are other items that can be helpful in getting the game on the air in the best sounding way. Take a small flashlight to be used when going into areas of little or no light, such as under bleachers or in storage rooms. It may also be used if the lights are dimmed or turned off for the introduction of the players. A calculator is small and convenient for keeping and figuring statistics. Binoculars are helpful during football games to identify players and follow the ball. The wide-angle lens works the best. Take along colored pens or markers and a bottle of correction fluid to correct any mistakes. For outdoor sports, take along plastic trash bags. They make great rainwear and can keep you and the equipment dry and functioning. They can also be used to carry equipment in an emergency.

Take along anything and everything that can possibly be useful in getting and keeping the game on the air. Always pack two of everything and even more than two of selected items. The more ways you can back up your equipment, the better the odds are you will keep the game on the air. Redundancy is important. Little items like equipment fuses can save a broadcast. Pack two kits and take one into the remote and leave the other in your car or where you can get to it on short notice.

GETTING THE BROADCAST ON THE AIR

One of the major challenges for a play-by-play team is to just get the game on the air. The famous (or infamous) Murphy's Law, which states, "If it can go wrong, it will and at the worst possible moment," says it all. When you go out of the controlled environment of the studio for a remote broadcast, just getting on the air can be a major challenge. Keeping the signal on the air can be just as much trouble. It may be that you have an engineer along to take care of any problems, but then again, you may be on your own. Make sure you cover yourself in as many ways as you possibly can. It all comes down to one thing: IF IT'S NOT ON THE AIR, IT'S WORTHLESS!

The lack of a broadcast will hurt your credibility, income, image, and sponsor commitment to your sports programming.

There are five basic ways of getting a remote broadcast signal back to the station for transmission. They are the telephone dial-up, a dedicated line (also called a broadcast loop), a remote transmitter unit (RTU), a satellite uplink and a cellular phone.

The Telephone Dial-Up

The telephone dial-up is ordering a business phone installation at the site of the game or event. You pay for the installation of the line, a monthly fee, and disconnect charge unless you leave the line in place. You need a voice coupler that will connect your mixer/amplifier to the telephone, unless your equipment has a built-in phone. Dial the station telephone number like any other call, set your levels, and away you go. This system has the advantage of letting you have a conversation with the person at the main studio. It is the best system for long-distance use in cost and practicality. It can have an excellent sound or one with plenty of snap, crackle, and pop. If you get a bad connection, redial and try again. If it still doesn't sound good, contact an operator and ask for help.

The input from any phone line should be run through an equalizer at the main studio. You can filter out and enhance the sound as you desire. The people at the scene of the remote can help the engineer at the studio set up the equalizer correctly by taking a sweep generator along. The generator sends selected or rotating tones/frequencies down the telephone line for the engineer to set the proper levels back at the studio. Another piece of equipment that can boost the quality of the sound is a frequency extender. Telephone lines operate on a rather narrow frequency. The frequency extender takes the lower-end frequencies usually left out of the transmission and mixes them into the phone line to be decoded at the receiving end by a second frequency extender. This gives the sound a depth and quality that would not exist without this equipment. The use of three phone lines with two different sets of extenders can be used to carry both the high- and low-end frequencies that are usually left out of the traditional telephone line. For the sound quality, the sweep generator and the frequency extenders are not very expensive for what they provide.

A Dedicated Line or Broadcast Loop

Another delivery system that comes from the telephone company is the dedicated line, also called a broadcast loop. The loop is a preconnected phone line that the phone company connects from the remote site to the main studio. The quality is usually very good. You pay for this line based on the time you use it and the distance the signal must travel. The further the signal must travel, the more it will cost. It may be more or less expensive than the dial-up system, depending upon the distance and the desired use. Unless extra equipment is installed, there may be no talk-back feature that enables the people at the remote site to talk to the people in the main studio. You may have to find a pay phone or other available telephone and call back to the station to check if everything is OK and the levels are all right. Always give the control room person a phone number where they can reach you, or at least be able to relay a message to you, if needed.

Remote Transmitter Unit

A third delivery system is the portable radio transmitter unit called a remote transmitter unit (RTU). The RTU can save your station a fortune in telephone company line charges and also give you portability. You can go where you want, when you want, as long as it's in range of the equipment. Because of the high cost of telephones, the investment in this type of equipment can quickly be recovered and justified. These units vary in cost, depending on your needs and intended use, starting at approximately $3000 and up. The RTUs are one-way transmission systems and do not have a talk-back system, unless it is installed as an extra or available on the more expensive models. The range of the signal depends on the terrain and the condition under which the broadcast is made. The units are usually good for 15 to 20 miles of coverage. It is possible to use more than one unit in a relay system to double your distance. If a station truck is available, an antenna stand that can be raised and lowered can add even more range to the line of sight unit. The quality is usually good and there is not very much interference. The system is simple and easy to operate. It is the perfect system for short notice work and will still provide a quality audio sound.

Satellite Uplink

The three systems already mentioned are the most commonly used in radio sportscasting, due to the cost and practicality. The use of the satellite uplink system is gaining in popularity with sportscasters who need and can afford it. Satellite use is expensive and requires an investment that is beyond many stations. You can rent the equipment needed and the time on the satellite from various companies. You need a ground station to send your signal to the designated satellite on the correct frequency. The satellite transponder (small transmitter/receiver) then receive your signal and rebroadcasts it to your station. The quality is excellent and the further away from the home station you are, the more cost effective the system can be. Many major league teams are now using satellites to send their away-game signals back to their flagship stations for rebroadcast. Some stations and networks are using satellites to distribute the games to member stations in the network.

Cellular Phones

A recent trend is the use of cellular phones. A cell, or district, is an area covered by a transmitter/receiver that handles mobile telephone calls. In metropolitan areas, this type of transmission is becoming very popular. This system can be used very effectively by radio stations for remote broadcasts, news, and sales. Cellular phones are still expensive and available for use in limited areas. If it is available in your area, it can provide a truly mobile system of communication at a reasonable cost.

DELIVERY SYSTEM COSTS

The biggest cost of sports game coverage is usually that of the delivery system. Here are several possible ways to avoid large costs due to the delivery system. Get the school to run an extension of their telephone line into the gym or field area press box. Your station should be very willing to pay for the extension costs because they will be far cheaper than installing and paying for a new telephone line. Make sure that the extension they are letting you use will not be accessible to anyone else during the time you need it for your broadcast.

Work out a deal with the schools/teams/stations in your league or coverage area. If everyone will put in a courtesy line for use by the visiting station, each station would only need to put in two lines for the regular season. Use the visitors' line and call your station collect, or give your station the number and have them call you. If you have a WATS line, the savings will be even more substantial.

If there is a telephone near the game site, in the press box, office, or even a pay phone, you may be able to hook into the system and avoid any installation charges. If you can get permission to use the telephone, the hookup is very simple. If it is in a multiline telephone system, you will have to run your signal through the telephone mouthpiece. This procedure is described in the equipment section. You should know where any telephones are in case of an emergency or equipment failure. Always carry enough extra equipment and cords to meet any possible emergency situations.

It may be cheaper and more practical, depending upon the situation, to record the game at the scene and then take it back to the studio for broadcast. With a good cassette recorder and quality tapes that receive the output from the mixer, the quality should be excellent. If the game or event runs at an inappropriate time for your station, or you are unable to get a line into the game or event scene, this may be the most practical route. This method is perfect for daytime AM radio stations which want coverage of the game from the night before. In any case, you should record every game broadcast at the scene as a safety valve in case the line fails, the signal is lost, or the station has transmission problems.

11

Season and Game Preparation

The key to a successful season is *preparation*. You must be prepared to handle any and all details to get the job done right. It can be a real strain to get ready emotionally, physically, and mentally in addition to the equipment and materials you must handle to get the job done right. The "details" are the difference between success and failure. By the time you get to the actual play-by-play of the game or event, the majority of the work is already done.

PRE-SEASON PLANNING

To get ready for the season, planning should start months ahead of the actual season or event. Management, sales, and programming must get together and set up the policies and planning details. Success must start from the top. Management must be ready, willing, and able to make a commitment to have a successful sports program. The cost of the games or events must be determined, plus the number of commercial breaks in each game or event that will be available, and the profit the station expects to make. Engineering must determine the equipment that will be needed and the cost of the delivery system. Programming looks at how to best fit the games into the station format and programming while the sales department figures how to best package and sell the games to sponsors. Radio and television stations use the same basic procedures to get ready.

A complete format must be written up so everyone involved will know what is going on and what is expected. The more contingencies that can be covered and backup plans formulated, the better the chances of success. Schedules of teams to be covered must be obtained and a potential list of games determined. The promotions for the season must be written and produced long enough before the season begins to build fan and sponsor interest. The sales department needs a minimum of two months prior to the season to cover their accounts properly. Some accounts are on yearly budgets and approximate prices must be available to accommodate them. Agency accounts may take even more lead time before they will reach a decision. Permission from the teams and schools involved must be obtained and any rights fees agreed upon.

PRODUCING FORMAT ELEMENTS

As the radio sports season draws near, the game open and close announcements must be produced. All the promotional materials for the season and each game must be put on the air. The commercials that are to run in the game should be produced, labeled, and placed separate from the regular commercials that are running on the station, even it if is the same as the regularly scheduled commercial copy. Backup copy and related production needs must be put on file and be available in case of failure of the recorded material. The program logs must be written to cover the games and supplemental shows. The board operators must be instructed in the game format and what their responsibilities are before, during, and after the game.

If lines or loops are to be used, they must be ordered a minimum of 2 ½ – 3 weeks prior to the game. Most telephone companies require 10 working days to make an installation. Try to work out a better deal with your local company. One very important lesson is to never trust the phone company. A day late for the phone installation is a game missed by your station. Make sure that all telephone numbers and names of people who might be needed in case of a problem are available. Always get the name of the installer of the phone, a trouble number that will get you in touch with a supervisor on duty at the phone company, and any installation numbers or order numbers that the phone company might need to effect a repair.

CONTINGENCY PLANS

Planning a game broadcast is much like a military operation. Contingency plans should be written for every eventuality that can be imagined. Play the "what happens if" game before they happen to you. You may be able to save ball games and revenues if you have prepared for most eventualities. The more confident about trouble and unexpected situations you feel, the better you will handle the trouble and the emotional and psychological upheaval that goes with it. If your phone line is disconnected during the game, who calls back? What happens if the game is rained out or delayed? The better the planning and the instructions to handle the problem, the better are your chances of success.

RIGHTS AND FEES

The ownership or rights to a game or event are very important to the sportscaster. For high school events, you must obtain permission to broadcast their games, usually by asking the principal. You should never pay a rights fee for a regular season high school sporting event. Your station is giving free publicity to the sports programs and generates a positive public image of the school system. The station also gives free time for school closings, late buses, lunch menus, and other school events. Radio coverage of sporting events, and the accompanying publicity, helps build audience interest and attendance.

At the college sports level, it depends on the market value of the team and the sport. If it is a major college football or basketball program, such as Ohio State, UCLA, or Oklahoma, the coverage will generate a major audience and money to match. If it is a small school or one where sports are de-emphasized, they should be happy just to have you there to give them the free publicity and support. Each situation and the money involved must be judged by its own merits. Make sure that you have written permission from the appropriate officials for any game or event.

For tournament events, there is usually a group or association that takes care of all rights fees and broadcast arrangements. This group, such as a state high school athletic association, will set the rules and determine the rates to allow the broadcast of a game or event. Those fees can go a long way when determining the profitability of carrying a game or event. Some stations will get together to form a network to split the cost and still cover the event.

Write a letter to the schools prior to the season requesting permission to cover their games, then call the school or team officials at least two weeks prior to the game or event and let them know you are coming and would appreciate their help. A little courtesy goes a long way toward a cordial working relationship. You can also check with them about any problems that might exist at that time. Make sure you have any parking and game passes or tickets to get into the game without any problems. If the RTU is to be used, check and make sure the signal will be of usable quality. All equipment must be gathered and checked before using it. Know what facilities are going to be available and if anything special or extra will be needed to handle the situation. Before you start out, check everything again. A written checklist can be a game saver. A forgotten piece of equipment might bring disaster. If you are going to use a pregame and/or halftime show, make sure it is located, labeled, and ready to roll. Label everything to avoid any mistakes or confusion.

ASSESS PRODUCTION NEEDS

Any production that can be done in the studio should provide the highest level of audio quality possible. Every part of the broadcast sound adds up to the total impression made on the audience. Use every edge you have to make a good impression on the listener. Quality is important to any radio or television station. Make your sports programs fit the station image or format. If you have a country music station, use appropriate music for the open and close of the game and any other time possible. Music attracts the audience attention and adds to their enjoyment. It makes your sound more memorable for the audience and helps promote your station format or sound.

GOOD COMMUNICATION IS ESSENTIAL

Even though we are in the communications industry, there is a major lack of communication on a person-to-person basis. Make sure that everyone involved (especially management) knows what is going on. Everything should be written out and explained to people who are connected with the broadcast. All format details, commercial breaks, shows, outcues, and every contingency should be carefully and fully explained. It takes a team effort to do a quality professional broadcast on radio and television.

Write out all the openings and closings and any other "intros" and "outros." This way you can enter and exit from the game with ease, confidence, and a professional polish. The opening sets the tone of the game. If you stumble and fall at the beginning of the game, you may struggle to get your performance and the audience attention at a good level, instead of building on a solid opening. The written close allows you to give everyone credit for the broadcast and avoid any hurt feelings. You can say "thanks" publicly to the teams, sponsors, and anyone concerned with the broadcast. Make sure you have a list of all sponsors so you can give them name mentions during the game.

Do as much paperwork ahead of time as possible. If you can fill out your scorebook or information sheet ahead of time, it will free you up to handle any problems that might occur. This also gives you a cushion to get yourself mentally and emotionally prepared for the game. Plan your travel time so you reach the game site a *minimum* of 1 hour before the start of the game. Depending on the circumstances and problems of the situation, you may need to be at the game site even earlier. As soon as you get there, set up your equipment and the delivery system, and make sure everything is working correctly. If it doesn't work, take it down and try again. Be patient and take it one step at a time. Correct what you can and then consult your panic list of names and numbers to call the appropriate person for help.

In case of doubt, write it out. Create a set of written notes during the game. These notes are valuable in jogging your short-term memory about specific plays or moments of interest. Create a form of shorthand or note-taking so you can write out information in the brief time that may be available and not distract you from the broadcast. In football and baseball, write out all scoring plays in detail. Written cues are valuable in providing in-depth information for later use for the audience.

PERSONAL PREPARATION

Through this entire book, it has been pointed out that the better prepared you are, the more likely you are to succeed. The more you know about the players, teams, situations, and the game or event, the more confident you will feel. That confidence will come across to the listening audience. You will sound more positive and comfortable in sharing the materials with the audience. The more materials you have prepared, the easier it is to ad-lib and fill time effectively in an entertaining manner. Your comments will have more depth and impact on the audience.

Once the equipment is set up and ready to go, get yourself ready.

Before any game, you must prepare yourself emotionally, mentally, and physically. Make sure you go to the bathroom, get a drink, and do anything else you need to do. You will be sitting or standing for an extended period of time and it's hard to leave once the game or event is underway. If you are uncomfortable, it is hard to concentrate on your work and do a good job. Take several moments prior to the start of the broadcast to loosen you muscles and vocal cords. Roll your head from side to side, tense and loosen face, neck, and shoulder muscles. Warm up your vocal cords by humming or singing and avoid damaging your voice. The voice must be properly warmed up and ready to go at full volume from the start of the game. Be like a cat that is relaxed, but ready to spring into action at any moment. Get mentally ready for the broadcast. Have your thoughts and attention focused on the game or event. Psych yourself up, have your emotional adrenaline flowing. It may help to give yourself a short pep talk to get fired up. You are a performer who must be ready to deliver your best when you go on the air. Run some plays through your mind and cover a list of descriptive words and phrases. Do a final check on everything. Come out ready to grab the audience's attention and keep it. Don't come out too hot and charged up. Calculate how much emotion and energy you need to catch and keep your audience. Leave room to increase the energy and excitement levels of the game as it progresses. Have a mental game plan prepared and then adjust that plan to fit the situation as it is warranted. Get mentally ready and emotionally psyched to give the audience a good show.

12

Radio Play-by-Play

A good play-by-play announcer provides a descriptive account of what is happening so the listener forms a mental picture of what is taking place. You will be working in the "theater of the mind." Work on the imagination of the audience. Paint a picture in your audience's mind with descriptive words as the colors, and your voice as the brush. The canvas will only be as colorful and exciting as you paint it.

Your voice is the game to the listener. Don't use a phony, so-called radio voice. Radio is a very personal medium. You are in a one-on-one situation with your listeners. The delivery, voice coloration, skills, and personal talent all make an impact on your audience. The better they like your voice and style, the more rapport they will have for your descriptions. How the audience perceives you, through your voice and delivery, determines your entertainment value and credibility. If you have distracting voice qualities, make them a part of your show or learn to eliminate them. Take any disadvantages you have and turn them around, if at all possible. Some things you just have to live with. Minimize any problem areas and polish your best qualities. Record every game and critique yourself and your performance. Have others listen and offer constructive criticism. They may hear things that are not apparent to you.

Learning to do play-by-play takes time, effort, and practice. The more you practice and work on your play-by-play, the better you should get. Start by simplifying what you see going on. Cut the action down to the most basic description of what is happening. Then add material as the time and the situation permit. In the three sports we will be covering, football, baseball, and basketball, there are three basic steps to doing play-by-play: setting the scene, describing the action, and giving the results.

SET THE SCENE

The audience needs to form a mental picture of the situation in the game. They will follow the play description much more easily and with greater enjoyment if you give them a firm starting ground. Prime the audience to accept your description of the action. With a solid mental starting point, the audience can keep up with and understand the play clearly, and be more involved and interested. Stop-and-go sports lend themselves to setting the scene prior to the action. Even in basketball, the situation must be set as the time allows.

DESCRIBE THE ACTION

Keeping up with the action is vital. If you get too far behind the play action, you are in trouble. Cut your words to the bare minimum, adding words and phrases as time permits and the situation needs. Don't get bogged down in too many details that aren't vital to the understanding of the audience. Get the most out of the words and phrases you are using. Keep all the action in the present tense. Avoid words like "and," "I think," and so on. They don't add to the play and they slow down your information flow to the audience.

Don't try to keep up with all the action that occurs in a play as it happens. Stay a half step behind the action. This will allow you to analyze and interpret the play so you can select what should be passed on verbally to the audience. You can filter out unnecessary materials and allow time for your emotional response to the situation. The audience cannot see what is happening so give yourself room to maneuver. Watch the play, evaluate what is happening, and vocalize it. Don't get too far behind the action or you may get lost or sound foolish. Skip over the unneeded details and keep up with the action. You cannot cover every little detail that happens; condense the action, giving the details of the contest. Make sure the audience has the material they need to visualize what is happening.

The selection of words you make during the play-by-play is of critical importance. Work to make each word have a high content/entertainment value. Try to be as accurate as possible with each word. Look for the fine nuances of the situation. Screen out excessive detail and focus on the moves or actions that individually constitute the play. Pick the descriptive words or phrases that best fit the action. Use a style that gives the highest word value of content and entertainment.

When doing a play-by-play, you don't need to use complete sentences. Use a shorter, telegraphic form of descriptive words and phrases. You must keep up with the action as it takes place. The audience cannot handle too much detail and you may get behind the action as it takes place. Cut the action down to the bare minimum of words. Use the minimum number of words that will accurately describe the situation. More details may be added as the situation allows. Make each phrase or word do the work of several words or phrases. Pick words and phrases that fit together to build a "flow" into the game. String words and phrases together that supplement each other and sound compatible or part of the same show. Work at a level that will attract your target audience and as many others as possible. Don't get into a jerky, stop-and-start situation with your delivery. Be a storyteller. Make the game understandable, interesting, and exciting for the listening audience.

GIVE THE RESULTS

After the play/action has finished, give further details and a summary of what has taken place. By giving the play a second time, you allow the audience to visualize anything that was unclear to them after the first description. It allows you the chance to cover areas that may have been secondary to the main action, but that are needed for further enjoyment or comprehension. Football and baseball offer major chances to cover the play or action a second time. During the recapitulation of the play, add a new twist or angle that will be of interest to the audience. If you use a very technical description of the play the first time, simplify the wording for the audience, who may not catch or understand it from the first time around. The better you know your target audience, the better you can angle the information for their entertainment. If you are working several audience types or groups, work on different levels when you summarize the action after the play.

MEMORIZE THE PLAYERS

One of the hardest things to learn is how to memorize the players involved in the action. If you cover a team regularly, you become familiar with their names, numbers, and faces and only have to learn the other team. In football and baseball, where they are positioned is a key to the player's identity. It takes time to memorize the people, and their numbers, who will be involved in the game or event. In football, due to the distance from the playing field and the difficulty in recognizing individuals in full football uniform, the numbers and positions of the players become vital to identification. In basketball, you must know all ten players on the court, even though certain players will be mentioned more than others. Look for physical characteristics such as hair color, glasses, braces, size, and so on that can be matched with the name and number of the player. In baseball, the position of the player and the batting order combined with the slowness of the game make it fairly easy to get the correct name and identification. Make a chart of the defensive baseball positions and insert the players' names. Glance at the chart when the action occurs to identify the player involved. Worry only about the last name of the players and match it up to their number or position. The sooner you have the names under control, the better you will sound.

It's important to focus on what is vital by using selective attention. The mind can handle information one area at a time. We sample information around us (called stimuli), and then select what we want to remember, and then store it in our short-term memory. The short-term memory can store from five to seven bits of information, each lasting up to 20 seconds. To retain the information for a longer period of time, it must be transferred into the long-term memory. Information can be moved into the long-term memory through the use of rehearsal. By going over the material several times while trying to relate the material to what is in our long-term memory storage, we can place the material where we can use it. The closer the new material relates to the information we already possess, the easier it is to retain the desired material. It should be noted that the use of verbal and physical activity can often help in retention of information. This would include reading the material out loud or walking while rehearsing. It may be helpful to make a game out of it.

The information we want to remember is often complex, and must be simplified and organized into separate "clumps" of information, especially when each part of the clump helps us to remember other parts. Categorize, or group together, information in some sort of logical order to make it easier to remember. When working with names, the "Ridiculous Association" technique works very well. The key is to imagine associations that involve odd substitutions or exaggerations, preferably combined with action. Make up sounds or phrases for names and then associate these words with a distinctive feature of the individual. For example: if a player's name is Duckburn, imagine a duck with a burning tail running with the ball. If the player is a fast runner, it fits the image. If he's a slow runner, the incongruity or counterimage may work.

We retain specific information when it comes to our attention and we focus on it. There is enough stimuli being sampled to allow us to gain a background or secondary impression. We focus on specifics, gaining a general impression at the same time. If we don't have enough information, we automatically provide it based on our own experiences and information we possess. We fill in gaps or areas we cannot observe by assuming things. It is important to really *look* and observe details.

There will be times when a list of names and numbers will *not* be available until just prior to game time. It's important that the play-by-play person takes the time to do the memory work. This time must be used effectively to enjoy a successful game broadcast. If the time to memorize is not available or very limited, start the game working the team name and nickname while identifying the key individuals' names and numbers. Work into using the names more and more as they become more familiar during the progress of the game.

If you cover a team on a regular basis, the work gets easier as the season progresses. After several games, you know your team and only have to work on memorizing the competition. The more comfortable you are with the names, the more confident you sound to your audience. Then you can concentrate on delivering the play-by-play descriptions. Build a memory file on each player. When they do something worth remembering, make a mental note of the action. Build a mental file on the teams that will prove to be valuable when filling time or ad-libbing later in the game or doing a wrap-up after the game.

Write yourself notes or cue cards to help your short-term memory. Work out a personal shorthand or note-taking system that will provide the maximum amount of information in the shortest amount of time. It must be simple and easy to use or it will not be useful. On major plays or moments of importance, write what happened in more detail. These notes are perfect for game wrap-ups, coaches interviews, and for use in the game itself. Give your memory plenty of help to improve your recall. The written cues allow you to accurately recall events and pass along the information to your audience for their edification and enjoyment.

REFERENCE POINTS

Have clearly defined points of reference or locations on the field or court that can be given to the audience. They need to know "where" you are talking about to base their mental pictures of the action. The audience must be able to relate mentally to the area you are describing to them. Make up a chart or picture with all the dimensions marked and areas described. List as many variations of a given area as possible on the chart. It adds to the descriptive variations you can offer to the audience. Work from a fixed left and right, and near and far on the court or field. There are ways to say "left" and "right" so the audience can relate, such as "moving left to right on your radio dial." In basketball and football, there is the team left and right side. If you are positioned at an end of the court or field, team left and right are different from your own when they face you. In basketball, know the dimensions in feet. Many times, a "10-footer from the right side" will be used to describe the location of a shot. Have the dimensions down accurately. Some sportscasters tend to exaggerate the distances.

BASKETBALL PLAY-BY-PLAY

Basketball is a *flow* sport. The action is basically continuous. There is little time to stop and analyze what is going on. To learn to do basketball play-by-play, start by describing where the ball is on the court and where it goes (top of the circle to the left wing to the baseline left). Don't use any names or identifiers yet. When you are comfortable describing the ball location, add how the ball is being moved (bounce pass, lob, chest pass). Next add the team name or nickname to the first two steps. This idea of the team name or nickname is very useful in a situation (usually the beginning of the game) when you are uncertain of the player's identity. This also helps the audience who may not know the names of the players. When you do all three steps comfortably, add in the name of the individuals. Normally you use the last name for the sake of time. Practice one play at a time until you can string them together. Then put everything into the context of the game, working with the time and the score. Don't try to cover every little detail. Hit the important points that the audience needs to picture the action.

FOOTBALL PLAY-BY-PLAY

In football, a play starts approximately every 25 seconds. The action of each play is about 5 or 6 seconds long. Football is a stop-and-go style of sport. The action occurs in spurts. Because of this time frame, football play-by-play has three distinct areas: setting the scene, describing the action, and then recapping or explaining the play when it is over. Be as technical as you want, but make it concise and descriptive. Make sure you cut your word usage to the minimum and then fill in as the time and situation permits.

Before the play begins, the audience must have the information required to picture in their minds the play situation: the location of the ball on the field, the down and the yardage required, plus the formation or offensive setup. If time allows, the defensive alignment should be included. Add any details that will help the listener picture what is happening. You are the eyes of the audience. One problem may be locating the correct yard line to give the position of the ball. This is very common on high school and smaller college fields that are not very well marked. If it is a problem, find any available and identifiable line and then count to the nearest line and estimate the distance to the ball. You may be able to look for the chain gang and ball marker on the sideline to establish the position of the ball.

When the ball is snapped and the play is underway, follow the ball. Focus your attention on the ball and what is happening just ahead of it. Don't worry about the defense until the tackle is being made. First, describe where the ball goes and how it gets there (pass, run, scramble). The more descriptive words you have in your vocabulary, the more entertaining and accurate your words will be to the audience.

Next, include the team name with the movement of the ball. Then add how the play was completed (out of bounds, tackled). Next, add the name of the positions that are involved in the play (tailback, split end, linebacker). Then you can go to the name (usually just their last) that goes with the position involved in the play. Put them all together and work one play at a time until you can string them together in a related, cohesive fashion.

After the play is completed, explain or expand on the details that you didn't have time to cover when the play was happening. You can describe the play in a different way or angle for further audience understanding. This is the perfect time for the expertise of the color analysts to shine through. They should use their expertise to explain and put the play into perspective. This delay is perfect for television and the use of the instant replay.

Memorize the people who handle the ball. This keeps the numbers you must mem-orize down to a manageable level. You can always look up the offensive line-men and the defenders as time permits. You will normally remember the outstanding defensive and offensive players anyway. Know about the team and its situation so you can keep the game in perspective for the audience. Football is a statistical game and the better the statistics you keep, the better you can explain to the audience what is going on.

BASEBALL PLAY-BY-PLAY

Baseball can be one of the harder sports to cover when doing play-by-play. The action is not constant. Baseball is a stop-and-go sport like football. The game moves along with minimal action and then explodes. The job is to keep the minimal-action times interesting and then keep up with the action when it occurs. This is one of the few sports where you do not always follow the ball. When there are base runners, split your attention between the runners, the path of the ball, and the defenders. When there are no base runners, you can concentrate on the ball movement and the defensive efforts. The location of the baseball is very important to the play. Don't hesitate to split the action into small, useable segments. For example, "The ball is hit into the gap between first and second, the runner rounds first and heads for second, the right fielder snags the ball and throws it toward second, a headfirst dive to the base, the tag is late and the runner is safe." Keep the action in understandable segments so the audience can picture what is going on and not get lost. You can always recap the action following the play and cover what you might have missed the first time. Don't get carried away and try to cover too much at one time. There are plenty of pauses that can be filled with information.

Baseball is basically a duel between the pitcher and the batter. A little hostility between them can help increase the audience interest, if you keep it in those terms. The pitcher is the key person on the field. Build a large vocabulary of terms and descriptive words or phrases that cover the pitches and delivery. The more variety you can offer the audience, the more interest you can generate. This also makes you more accurate in your description of the game. Give the count after every pitch to set the scene for the audience. Time is in your favor in this slower moving game. Take advantage of that fact.

After you set the scene by giving the count and position of any base runners, set the defensive alignment, the stance of the batter, and other details as time permits. If the tactical situation calls for a specific defensive setup, point it out to the audience. You should be expert enough to point out any special defense or offensive strategies and situations, such as the infield moving in in case of a bunt or to set up a double play. If you explain the possible options to the audience, they will be mentally "geared" or ready to accept the play you are calling.

Baseball is a very statistical game. Make sure you keep a complete scorebook and information sheet. Baseball is usually slow enough to allow adequate time to cover the game information. A regular baseball scorebook may be adequate for your needs, or you may want more room to keep up with the information you think the audience needs. Use 3 x 5 note cards or a sheet of paper and list the defensive players in their positions. If the ball is then hit into left field, you look at your chart and see who is there. If you don't have time to look for the name, use the defensive position for identification.

BUILD THE ACTION

Each game has its own high point of action, or climax. Build the audience up to that point in the game, just as a good movie or book does. Set the groundwork and build up the suspense to get the maximum audience attention and involvement. The first quarter or early innings should set the scene, introduce the characters, and show the conflict and strategies that are being used. In the second quarter, build up toward the halftime break. It's like a mini climax that temporarily satisfies the audience, yet gets them back for more in the second half of the game. Try to leave them at a good energy level. In the third quarter, reestablish the player identities and situation. Start building momentum toward the coming climax. By the fourth quarter, provided the climax isn't already over, build toward the high point. Increase the energy and volume levels. Give the high point, or climax, the drama that it deserves. After the climax, help the audience come down and enjoy their moment of satisfaction. Give them the emotional release you built up during the entire game.

It may be difficult to keep your attention where it belongs during a game. There are many things going on that can distract you. You are the eyes of the audience and need selective tunnel vision. The audience needs a continuous flow of information to stay in the game and enjoy it. There are times when you must split your attention among several things. You may have to watch a play, make a mental or physical note, describe the action, and plan for the next movement.

TERMINOLOGY

Announcers must know and use the terms for each sport. Each sport has its own terminology and sayings. You assume some audience knowledge of the terms of the sport. You need to build up two large vocabularies. One of the sport terminology and one of descriptive terms and phrases for use in any situation. Use a thesaurus to find as many words as possible to cover or describe a situation or action. The more variety you can provide the audience, the more interesting it will be. Variety is the spice of life. Take a list of words and terms with you to the game. Use that list to supplement your normal vocabulary. The larger your vocabulary, the more chance you have of fitting the right words to the situation. If you discover you are using the same words or phrases too much, write out the word or phrase in big, bold letters and place it in front of you during the game as a reminder not to use it. You may invent your own terms or phrases to fit a specific situation. You are an entertainer in the theater of the mind.

TIME AND SCORES

The audience always wants to know how much time remains in the game and the score. Except for games where the winner has already been determined, the time and score create the drama of the contest. It is frustrating for the audience waiting to hear the score of a game not to hear it. The radio sports audience tunes in and out of the game broadcast. While you want them to listen as much as possible, it must be balanced against their needs. The closer the score is, and the less game time remaining, the more you should give the time and score. The further apart the score, the less you give it to the audience.

Football is a sport for which it is easy to forget when to give the score. Work on the habit of giving the score and time remaining on every change of possession or possibly every first and third down. The use of an egg timer or watch with a buzzer may also work. In basketball, every time a basket is made, or on a change of possession, the score and time should be given. Baseball offers logical breaks in the action, plus the slow pace allows plenty of opportunities to give the score and inning. The inning gives baseball the same time factor that the clock gives football or basketball.

USE AN ENTHUSIASTIC, POSITIVE DELIVERY

The speed of the announcer's delivery is an important asset. Use the speed of the delivery to show emphasis and promote excitement. The pace must be fast enough to avoid audience boredom and yet slow enough for audience understanding. Vary the speed of the delivery to fit the situation. You cannot keep the audience in a state of constant excitement or it soon becomes dull. Work your audience up and down a roller coaster of emotion and excitement. Use the pace of your delivery to build toward the climax of the game. A pause in your delivery can be used to point out items of significance and to build suspense. Some things are better left unsaid. Silence by the announcer can be golden. Let the crowd speak for itself. Crowd reaction shots on television are very common. The crowd reaction, or lack of it, adds depth to the situation for the audience. Give your audience and the game crowd a chance to enjoy the drama or emotion of the moment.

Try to hit the audience on an emotional "gut" level. If a player is reaching out to grab the ball and it goes past his fingertips, let the feeling, the emotion of the moment, come through in your voice. Get the audience emotionally involved in the game. Playing with their emotions gives the game more meaning to them. Be sure to give them both the good and the bad. Let them enjoy both the pleasure and the pain, but don't get carried away with the situation. Don't let things get too personal, keep it at a professional level. To a certain extent, it must be a real emotion or it comes across as fake.

If you don't sound enthusiastic and excited, the audience will not get excited either. Enthusiasm is contagious. If you are bored and sound that way, the listener will also be bored. Sports are exciting and this must be reflected in your voice. Have a good energy voice level when you start the broadcast. Don't come out too hot. Always leave yourself room to build with the excitement of the game. Don't peak out, because something even more exciting may be coming up. Keep yourself and your voice under control at all times. Don't start screaming and over-modulate the signal to where the audience cannot understand you. It must be exciting, but it must be under control. You are the eyes of the audience. If they can't understand you, the whole thing is a waste.

Sportscasters are entertainers. Most of the better jobs go to people who are more colorful and entertaining. Radio and television are a part of show business and must be treated that way. Sportscasting seems like a stepson to news, but it is artificial and should be considered entertainment. Put some pizzazz into your delivery and approach. A flashy approach goes along with doing a professional job. Make sure the style you cultivate fits your personality, the station, and the desired image. Don't be shy. If you don't toot your own horn, nobody else will.

If you find a gimmick or act that adds to your presentation and the enjoyment of the audience, go for it. If you don't experiment or try out new ideas, you may get stuck in a rut. If something doesn't work, write it off to experience and move on to something else. Build a catalog or file of successful ideas.

Humor is a double-edged sword that can cut both ways. Some humor can be a very effective tool and help relive the tension for both the announcers and the audience. This helps achieve an emotional balance. Have a point to any humor you try. Keep it appropriate to the situation. The fans, teams, and coaches are taking their game seriously, so avoid getting carried away with laughter, jokes, and stories. You may get your audience upset enough to drive them away or make them less receptive to your broadcast. What you think is funny may not be to others in your listening audience.

Keep the game broadcast as positive as you can. Set a positive tone at the beginning of the game and carry it through to the end. Don't get down on, or strongly criticize, the players, teams, coaches, fans, or officials. You do root and urge on your team, but don't badmouth or make disparaging remarks about the opponents or personnel involved. The audience likes it when you are rooting for their side. If your station represents both teams, it is vital not to show too much partiality. You may be accused of it anyway, especially by a losing team. Keep everything sounding positive. Let your voice coloration and tone take care of any criticism of a play or person. Be a "homer" in the sense that you are rooting for your team. Keep the level of the game on a professional tone and level and not like some amateur booster of the team.

13 Game Coverage Aspects

There are many areas of concern and details that make a game broadcast a success. While each situation will be different, here are some ideas that may help your play-by-play game coverage.

AUDIO CUTS

Recorded audio cuts of coaches, players, and related personnel can be used by radio stations during a sportscast or during the game itself. Place the audio cuts in the order you are going to use them with written intro and outro cues. They give an added dimension to your coverage. If the coach comments on a certain type of defense he expects to see, use it during the game at an appropriate time so the audience gets it first-hand. It sounds much more impressive and professional. Television sports have used prerecorded audio and video clips during games for years.

PLAY UP HEROES AND RIVALS

Sports are one of the few remaining areas where people can find heroes. People want and need people to look up to and emulate. If your audience wants a hero, give it to them. A sense of hero can be achieved through declarative statements and by voice coloration. Let the emotion and admiration come through in your voice. A believable buildup of a person is fine, if warranted. Do not get carried away with this idea or it can hurt your credibility. Be careful that building up a hero doesn't backfire if something goes wrong. Nothing crashes harder than someone who was on top. People who were the biggest admirers can turn into the most vicious critics if their illusions are crippled or destroyed.

Create a feeling that the competitors don't like each other. The big games are usually between schools or teams that don't seem to like each other very much. The emotional buildup in a big rivalry seems much bigger, and the fans and the teams both reflect it. Studies have shown that fans react better, and are more emotionally involved, if aggressive hostility is present. Fans seem to think if the competitors don't like each other, they put out more effort and try harder to win. Work the audience at the emotional level and get them involved in the game. Try not to take it down to the individual level very far. Keep it at the team spirit level. Keep the game in perspective and don't get carried away with it.

COVERING INJURIES

People are going to be injured in sports activities. They are a part of the sports scene. You must be ready to explain and handle any injury situation. *Do not speculate* on the extent of an injury. Get the facts before you speak. If you don't know, don't say. Keep the injury activity on the field or court where it happens. Don't get carried away with the activity. If you speculate or say more than you know, you open yourself up to criticism, a loss of credibility, and possibly even legal trouble. Plan ahead on how to best handle and fill around an injury situation. Try to keep everything in perspective. This is a judgment call on the part of the announcer. Keep your discussion to the basic visible facts and limit your speculations on the extent of the injury.

The more you know about sports medicine and injuries, the better you can explain the situation to the audience. This knowledge adds to your credibility when you give an authoritative presentation about specific injuries and their treatments. A knowledge of preventive techniques is great fill material and continues to add to your reputation as an expert on sports.

ONE-SIDED CONTESTS

Most play-by-play games you cover will not be the close, exciting contests you hope will take place. Those are the easy type to do. Some games are unexciting, boring contests where one side or the other clobbers the competition. You want to keep the audience attention for as long as possible and keep them and the sponsors happy. The best way to keep the audience is to limit giving the time and score, and concentrate on the immediate action that occurs. Take the game out of context and work on the game one play at a time. Make each and every play seem important. Even then, some games just will not be entertaining and hold the audience. Keep plugging away until it is over. There is little else you can do.

RULES AND OFFICIALS

There are good and not so good officials in sports. There are well-officiated games and poorly officiated games. The officials have a tough job and it is not the announcer's place to make judgment calls or second-guess the officials' decisions. *Don't* criticize the officials! They are going to make mistakes (they aren't perfect like announcers are). It is just part of the game. Everyone wants a well-officiated game, but it isn't going to happen all the time. The announcer's job is to report and describe the game to the audience, not to make value judgments on those officiating. How do you know it wasn't the correct call? The reverse angle used on television replays has shown how the angle of view makes the difference in the call. A good official will be in position to make the best call. You have one viewing angle, while the official has another. If the crowd doesn't like the call, mention it and let the crowd noise say so. If you don't think the call was correct, let the "tone coloration" in your voice express your feelings. Whether it is right or wrong, the official's calls are part of the game. Try to keep everything in perspective for your audience. It is fine to point out the effect the officials are having on a game, but not to directly criticize the actions of the officials. It usually takes more than one bad call to lose a ball game.

Take the rule book with you to every game. Even though you should know all the rules by heart, this allows you to quote, or clarify, the actual rule to the audience. It adds to your credibility with the audience and can be valuable fill material. In case of doubt about the ruling, find the rule and sound like the authority the audience expects you to be. Make sure you have the current rule book and have the up-to-date rules and any changes.

Make sure signals given by game officials are understood by the announcer to pass along to the audience. Even if you anticipate the call, get confirmation from the official. Wait on their call to avoid mistakes and confusion by the audience. If you speculate, tell the audience that is what you are doing. This is especially important in football, due to the distance from the playing field. Know which official to look at to get the correct call.

It is vital that the play-by-play person and partner have a visual set of signals planned out to avoid mistakes and confusion. Work out visual signals that indicate various needs or information during a game. The signals must be easy to understand and convenient for both partners. Have paper available to write out notes to each other if signals aren't enough. You need every edge you can find to create a smooth professional performance.

DELAYED GAMES

On radio, it's fine to record a second game and play it over the air at the conclusion of a first game or other programming. It should be considered as valuable as a live broadcast. This is one way to cover games or events that run at the same time or overlap. If you have a way to get both audio signals back to the station, record one game on reel-to-reel tape. Each situation will determine which game will be aired live or recorded for playback. Make sure that all tapes are clearly labeled and are being recorded correctly. The game announcers can time the commercial breaks so the tape can either be left running or cued up to the next starting point by the board operators. Announcers should remember that the tape is always running and comments that are made during a break may end up on-air. At the beginning of the game and the second half, make the announcement that "this game was prerecorded for broadcast at this time." Make sure there is time during a logical break to change the tapes. Check every break to make sure there are no problems.

It may be possible, and probably cheaper, to tape the game at the game site. Use a cassette recorder wired to the output of the mixer. Make sure there are enough tapes, labels, and batteries to do the entire game. Change tapes at logical places and clearly mark them. If you have a daytime station or are unable to secure a line from the game site, this is a low-cost practical way to cover sports. You should record every game at the scene with your cassette as a safety measure. If you should lose a line, the power goes off at the station, or any other problem, you still have the game in the can and ready to go when it becomes possible to put it on the air. If the game doesn't get on the air, it makes the listeners unhappy, the sponsors don't get their ads on, so no money is being made, and the station's credibility goes down the tubes.

TIME AND SCORE

The time that remains in the contest is the game time. The audience needs to know the time frame of the game. Whoever has more points when time runs out is the winner. Use the game time to your advantage. Build on the time remaining in the contest for suspense and drama. Work the clock and the score to build audience excitement. Work the time remaining to give your sponsors the best available impact on the listener. The more involved that the audience is with the game, the more receptive they will be to the commercials they hear. It's important to get the commercials on the air before the game runs out and the audience departs.

Getting into and out of a game on time can be very important. Will you end earlier or later than expected? Will you be joining a program in progress or does it matter? Keep an eye on the clock and be ready to wrap things up quickly or stall for an extended period of time. This is very important when doing a television game broadcast. Television programs usually start and stop in half-hour blocks. Radio tends to run in shorter time segments that are easier to join.

PERSONAL NEEDS

If you are going into crowded working conditions, make sure you arrive early enough to claim your working space (plus a little extra). You may have to guard your space against late arrivals. The extra margin of space you obtained lets you give up some room and yet keep all your working space. Be prepared to stand up for your rights and space. You may have to get pushy, or resist pushy people, to get what you need to do the job correctly.

If you are uncomfortable, it is hard to concentrate on the needs of a game broadcast. If you are forced to sit or stand for an extended period of time in one location, the temperature and conditions will seem more extreme. Dress for the weather or conditions you might encounter. When working in cool or cold conditions, dress for at least 20 degrees colder than anticipated. If it is windy, it will seem even more extreme. If hot and sunny, find or bring some shade with you. Always give yourself a margin of error in favor of your personal comfort. A seat cushion or pad can help tremendously when sitting in the bleachers or on a bench for an extended period of time.

Take care of all personal needs prior to a game or event. Make sure you are physically, mentally, and emotionally ready to perform. Take along something to drink during a game. Dress for the anticipated outdoor weather or indoor conditions. Remember that you are the representative of the station to the audience at the game. Look the part of the professional, but with an edge toward personal comfort.

Know where every emergency exit is located in case of fire or trouble. Find the best location to take shelter in case a fight or riot occurs. Announcers are usually among the last to leave the game site. Parking lots may be hazardous to your health if you aren't careful. Keep alert and avoid any potential trouble situations. With your hands full of expensive equipment, it could be difficult to protect yourself, your equipment, and your wallet. You could become the target of a hostile crowd. You are a visible representative of the other team and could be a target for abuse. It is a good idea to know where the nearest officials or security personnel are located. If it looks like trouble, ask for help. A crowd can turn into a mob very easily. Don't be foolish. Use your best judgment concerning your personal safety and take sensible precautions.

14

Television Play-by-Play and Color Commentary

AUDIO FOR VIDEO

Television sports play-by-play is different from radio play-by-play. Television play-by-play is like putting the captions on the pictures. It requires the announcer to explain, clarify, and supplement what the audience is seeing on their screen. The audience can see for themselves what is happening. The announcer is trying to provide information to the audience. Try to develop the personalities of the team and the players. Help the producer and director locate the story lines and any information that lets the audience feel they have inside information. The announcer must be aware of what the audience can or cannot see. Know the camera angles and areas of coverage. One major advantage of doing a television game is the instant replay. If you missed the action on a key play or event, you can catch it and analyze it the second time around. The replay is the perfect situation for the analysts to show their expertise. TV sports are perfect for the two-person announcer crew that most stations and networks use. As in radio, the working relationship between the announcers is important to present a smooth, entertaining show to the audience.

Work at building suspense and excitement through your vocal efforts, combined with the time and score. Try to sound confident and sure of yourself, even if you are encountering problems. Smooth over any problems that the audience might see, and support the efforts of the director. Provide basic names and situations for the audience. Set the scene and let the play run into it where the audience can see it. Have enough game and team knowledge to anticipate and plan for a specific style, and then adjust during the game to fit what actually happens. Build a mental file and use paper notes of the action that has taken place for later use. During a commercial time-out, any suggestions or shots you would like to see should be mentioned to the producer-director. It's important to have a large vocabulary of descriptive and technical terms. Give the audience some variety while describing the action and situation as clearly and informatively as possible in an entertaining manner.

LISTEN AND CONCENTRATE

Television play-by-play takes split concentration by the announcers. You must listen to the on-air delivery and to the calls over the intercom by the director of the telecast. The director is the person who controls the game for the television. The director calls for the camera shots to be used, the length of those shots, when replays will run, and any other features or breaks. The announcer must support the director with the audio portion of the game action. The announcer can focus the attention of the audience as the action takes place. Studies have shown that the vocal excitement and descriptions by the announcers add significantly to the enjoyment and understanding of the game by the audience. The picture carries the action and the announcers work on the audio-emotional aspect of the audience. The announcers and director work to build the game to the climax for the enjoyment of the audience.

Television play-by-play requires the announcer to listen both to the program output and to instructions from the director. Listening to the director and keeping a flow of information to the audience may be difficult. Listening to two different sources may be very distracting. Create mental awareness for key words and phrases. Learn to shift your attention and focus on the needs of the moment. The calls from the director that are important for the announcer include *replay* and *counting down* into a break. Practice talking over a countdown from ten to one, so you finish at two or one. The score, teams involved, and your station or network should be included. If there is something exciting or dramatic, it might be used to "tease" the audience so they come back after the break. A music "bumper" is used to fill time until the break gets underway.

Be aware of what is taking place around you, yet maintain an entertaining, nonstop flow of information to the audience. The director may ask for specific information, such as time and score. Work such material into your game commentary in a natural way. Practice toss-offs or hand-offs to another source, such as an interviewer or station anchor. Plan what you want to say and run through it mentally. If time permits, write cue notes as a guide.

THE GAME FLOW

Many of the techniques used by radio play-by-play announcers can be adapted and modified to work for the television announcer. With the game or event action being carried by the video, the announcer is free to concentrate on the audio excitement and emotion. The announcer is like a ringmaster at a circus, directing the attention of the audience from one exciting show to another. The announcer controls the continuity of the broadcast. A smooth flow of pictures and words in continuous action lets the audience get into a rhythm or flow of the game. When the audience gets into the game flow, they have a greater mental participation in the action and derive a greater feeling of satisfaction. This mental participation also gives the commercials in the game a greater impact with the audience, due to the higher attention level.

The opening for many games contains video of the competing teams. The announcer usually writes her own introduction to cover the opening shots and give the opening sponsor credits for the game. It's important to be ready to get in and out of commercial breaks and other segments smoothly and inside the time frame of the director. The announcer may also read cross plugs and promotional messages for the station and team.

WORKING WITH A PARTNER

The play-by-play announcer must work on different levels of the audience's emotion and intellect in coordination with his analyst. If the analyst is very emotional and strikes the audience on a certain emotional level, the announcer must work to cover areas that are not being covered. If the analyst is given to the anecdotal approach, the announcer works more on the facts and a straighter approach and appeal. If one way doesn't affect the audience, hopefully, the other way will. The announcer must keep the game in perspective for the audience. Because of the showmanship needed to catch and hold the attention of the audience, the announcer must blend his personality with the partner's personality and approach, the supply the audience with a large spectrum of appeals on both then emotional and intellectual levels.

It is very important that the play-by-play announcer be in charge. It is part of her or his job to control the tempo of the broadcast. It is very confusing for the audience if they are stepping on each other's lines and comments. Work out a set of visual signals to avoid crossing up each other. Speak one at a time. Try not to repeat exactly, or paraphrase, what was said by the other person. The color commentator doesn't have to speak at every break in the action unless he or she has something to add. Be prepared to make intelligent remarks at logical breaks in the game and at halftime. Be precise in the comments you are going to make. Generalities don't add very much to the show. Do the homework needed to fill in the facts and figures the play-by-play person doesn't use or may forget to use.

The color commentator must learn how to fill in during the pauses created by the play-by-play announcer, when it seems appropriate. Keep your comments short and to the point so you can give it back to the game announcer before the next play or action begins. Try to keep your talks friendly. Audiences prefer the conflict to be on the field and not in the booth. Try not to correct your partner on the air or at least try to keep it subtle. Pass a note or discreetly mention something to remind or correct him or her. Don't throw your partner a curve ball with a comment that comes out of nowhere and doesn't seem relevant to the situation. Keep your comments aimed at the audience. Work hard to complement your partner and provide the verbal support needed for a good broadcast.

Any play-by-play team must learn to work together in a coordinated manner that enhances the broadcast to the audience. Each person must work in his own area of responsibility. The person who is doing the actual play-by-play is in charge and controls the tempo of the broadcast. The play-by-play announcer must describe the action as it occurs in an entertaining manner. The color analyst must explain what has happened or is about to happen, or point out items of significance to the audience. Certain people seem to have the chemistry or personal characteristics that make them a smooth coordinated team. Some partners never seem comfortable with each other. A compatible team that has worked together seems to read each other's minds and know what each is going to do or wants to do. During a television broadcast, as the replay is taking place, the analyst usually describes the action and explains what is going on. The play-by-play announcer must keep an eye on the field or court to avoid missing anything as the action resumes, even if the replay is not completed.

In radio sportscasting, the announcers are the eyes of the viewer. In television, the eyes of the viewer are on you. You must be prepared to look and act the part of a professional television personality. Dress for the part and be carefully groomed to look good on camera. When you are on-camera, look into the lens. Make the camera lens a person that you are carrying on a one-way conversation with. If you are with a partner, talk to both your partner and to the camera, as if the camera is the third person in the conversation. Address your comments to the appropriate "person" as the situation requires. If you are talking to more than one person, talk to them one at a time and avoid any confusion. If you are holding the microphone, get it close enough to the mouth to insure a good quality sound, but without blocking or obscuring the person's face. A monitor may be nearby so you can see the camera angle and shot being shown to the audience.

15

Scoring and Identification

Scoring and identification forms for sports announcing need to provide easy visual access to scoring and player information. The form or style used needs to have the information in big bold letters and numbers. There are many commercial scorebooks and statistical forms available. If you find a form or style that works for you, use it. If you find you need more or different information than is provided, modify the forms to suit your needs. The form must provide easily accessible information during the game or event. The sports announcer needs *current* information, as opposed to a newspaper or magazine writer who works with the materials after the fact.

When writing the names of the players into the scoring or identification forms, spell the names as they sound, not as they are spelled. The announcer must worry about the pronunciation and not about the spelling. The announcer's only concern is taking care of the listening or viewing audience.

Use the forms in this book, buy commercially available forms and score sheets, or develop your own. Find out what works in your situation. Check with other sportscasters and statisticians to see what they use. Pick and choose among the many styles and forms that will provide you with easily visible and accessible information about the players and the game. The forms you choose and the style you use are important tools. The forms in Appendix A are described in the following sections on football, basketball, and baseball scoring.

FOOTBALL SCORING

In football, the announcer's main concern is with the offensive team. This form provides a visual depth chart of the offensive team plus areas for limited statistics. On the top line of the form goes the team names, record, and coach's name. The next two lines are for the offensive linemen (tackles, guards, centers). Numbers and last names should be boldly written and other information included where space is available. The large squares are used to identify the personnel who may handle the ball. Put the quarterbacks in the middle and the running backs, tight ends, and wide receivers in their appropriate areas. On the left side of the form, list the name, number, and any statistics for the punter or kicker. On the bottom left side of the form that reads "number of plays per series from start," write down the yard line where the team takes possession of the football. Then keep a running count of the number of plays in the drive before a score or change of possession. On the right side of the form are the areas to keep track of first downs, pass attempts and completions, pass interceptions, fumbles (kept and lost), and team penalties. A complete description of any scoring or plays of major importance should be kept on a separate sheet of paper, along with the quarter scores. When doing play-by-play or color commentary, this is an easy-to-use form. Defensive players can be picked out from a roster, separate sheet, or depth chart.

Along with individual identification of each offensive football player, team scoring and statistical information is needed. In the top quarter brackets, keep quarter scoring and total game scoring. Each team gets one side of the form. Each block will show the time, quarter, game score, and how the play occurred. Other statistics are crossed off to keep running game totals. Colored pens can be used to show in which quarter things occurred, or draw a line after each period is complete.

BASKETBALL SCORING

Basketball announcers need to keep up-to-date statistical information about the game and the individuals involved as it progresses. On the top line, put the team name and nickname, record, coach's name, and any other information that fits. The shot chart (numbers 1-99) is for the team shots taken in the game. At the end of a quarter or half, draw a vertical line to show the shots taken by the team in a particular period. Under the shot chart, you may want to display any ties that have occurred in the game. Just put the numbers that the teams tied at as they occur. In the blank space in the upper left corner, keep the difference in scoring between teams. You can say by how much each team has led in the game. As the turnovers occur, mark out the appropriate number. The numbers 1–25 in the upper right corner can be used for keeping count of the commercials played, rebounds, or other information that seems appropriate.

In the name column, always list the last name in big bold letters, next to the numbers. The first name, height, and other information can be included in smaller numbers and letters. It is easiest to put the players in numerical order (smallest to largest numbers). Some announcers find it easier to put the starters in the first five slots and then the rest of the team numerically. Never use alphabetical order for players' names on a score sheet. Use different colored markers or pens to show what quarter or period the shot, foul, or other event occurred.

This form is designed to keep a running total score for each player. Just add the points together as you move from left to right on the chart. To show foul shots on the chart, put a circle for each attempt. If it is a two-shot attempt, put one circle over the other. As each shot is made, fill in the circle. If the shot is missed, leave it open. Always mark the top circle as the first shot and the bottom circle as the second shot. If it is a foul shot to complete the three-point play, put the circle next to the number of points for the player earned on the play. If it is a one-and-one attempt, make a circle near the top and put a stem or line down from the bottom. If the first shot is made, fill in the circle and draw another circle connected to the stem to show the second shot. If it is missed, the stem line shows the attempt was made as the first shot of the one-and-one. Put the total points of each quarter or period in the correct space and add the scores from there. If there is a three-point shot, just add three points to the score and draw a circle around the points to indicate the three-point shot. If a player fouls out of the game, total his points and write the time at which the player had to leave the game. Put lines around the time and score to point it out visually. Player fouls are in the center of the page with the team fouls listed on the bottom of the page.

BASEBALL SCORING

Baseball is a stop-and-go sport that provides easy identification plus time to keep extensive statistics. When using a baseball scorebook, most commercial forms work very well. On the top line go the teams, nicknames, records, where the game is played, and when. The players are listed in their batting order of appearance. This type of scorebook can be used one of two ways. A line can be drawn through the appropriate abbreviation on the right side of each square to indicate what happened. A symbol can be drawn in the correct box to indicate what happened. Most commercial scorebooks work very well or can be adapted to suit your needs. In baseball and softball, the action is usually interspersed with long enough pauses to allow the keeping of an accurate scorebook. At the bottom of each inning column on the summary line, it always reads "runs, hits, errors, and left on base." The totals from the summary sections are then included as the line score.

Some of the abbreviations used include the K-strikeout, DP-double play, SB-stolen base, LOB-left on base, W-walk, HPB-hit by pitch, WP-wild pitch. Any play circled is an out. The F with a number means a fly ball to a person at that numbered position. A line with one line across it is a single. A line with two lines across it is a double, and three lines across is a triple. Put HR for any home runs that are hit and darken in the diamond. This is only one form of scorekeeping. Any style you find that works for you is the correct style.

A 3 x 5 note card may be attached to the scorebook so you have both the batters (offense) and the fielders (defense) in front of you. Put the position number next to the name for easier scoring. If the ball is hit into center field, glance at the card to and know who will be trying to make the catch. This same idea of picturing the names and positions of the defensive players can be done on a sheet of paper or notebook and laid beside the scorebook for easy access.

16

Producing and Directing Television Sports

WHO DOES WHAT?

The sights and sounds of a television sportscast are determined by the work of the producer and director. The producer is in overall control of the broadcast. The producer decides on the overall look of the broadcast by picking out the story lines and angle of approach, all based on his view of what the audience wants. The director handles the nuts and bolts of the operation. The director works the length of the camera shots, decides which replays will be shown, and makes other decisions about the actual broadcast. In many markets, the job is done by just one person, appropriately called a producer/director.

Let's look at how a typical situation might be handled. Management, sales, and programming decide they want to carry the local college basketball games, both home and away. A schedule of games is drawn up and contact made with the team. Rights fees for permission to carry the games must be negotiated. A tentative budget is drawn up. Equipment needs, personnel, delivery systems, and other details must be examined and their costs determined.

The negotiation of rights fees can be critical to the profitability of game broadcasts. The standard feeling among school or team officials is that television coverage hurts their gate (ticket sales). Some schools or teams will not allow local broadcast unless they are sold out (as done by the NFL). It may be the station might have to pick up the cost of the empty seats before the broadcast will be allowed. Officials hate to give up seating space for the placement of the cameras, even though they make money from the rights fees. Many school and team officials are now realizing the value of TV and radio coverage in developing a long-term loyal audience.

Once a preliminary budget has been approved, equipment, personnel, and the distribution system must be arranged. Many stations have their own production equipment capable of working a game broadcast. For away games, it's usually impractical to take your own equipment. A deal is usually made with a local television station or production company to supply the equipment. A crew to work the equipment must also be put under contract. The talent and the producer/director go to all the games to maintain continuity and insure a quality broadcast.

A vital element in a TV sportscast is the distribution system. It's how the video signal is sent back to the studio for transmission to the audience. There are three basic systems or ways to move the signal back to the studio. For local broadcasts, a microwave system is used. An antenna is aimed at the station's microwave receiver, usually from a remote truck. This system is good for limited distances, usually not more than 30 miles, depending upon the terrain and conditions. The satellite delivery system is the best system for long-distance games. You rent a satellite truck that will send the audio/video signal to a satellite in orbit. The satellite has a transponder that receives the signal and transmits it back to earth where it is picked up by a dish at the station. You rent time from the satellite owners and pay by the hour for its use. A delivery system commonly used is *Video Land Line* from the telephone company. Video Land Lines are used, but in a limited way, due to the use of satellites.

A common way to save money, when two stations want to carry the same game, is to split the signal feed. In a split feed, two separate signals or outputs are sent from the control truck to the separate stations that want the signal. The basic video action is the same for both stations, except the visiting station is given a small switcher and a designated camera for putting their own talent on the air at predetermined times. The audio engineer sends the correct audio signal to each station to supplement their video signal. The visiting station usually sends their own talent and a producer to look after their interest. ESPN and other networks use this formula at times. Since the costs are shared, it is cheaper for everyone concerned, and the crew makes more money for, in effect, doing two games at one time. In a shared feed, the announcers for one station go on the air, while the other station cuts away to its studio for commercials or other materials, until the first set of announcers are done, and then the process is repeated for the other station's benefit. It takes careful timing and planning by the director and the announcers to avoid any overlapping problems, and a sloppy looking broadcast.

SITE SURVEY

Before any equipment can be ordered for a game broadcast, a site survey must be carried out. The director, or someone who can examine an away site, must go to the game site and examine the physical layout. Where will the cameras be positioned? How much cable must be run to connect the cameras to the production truck? Where will the truck be parked? What delivery system will work best? Where are the announcers seated? Is there an interview area that can be lighted? These and many other questions must be answered before equipment can be ordered. Following the site survey, the appropriate equipment and personnel are contracted to work specific games. The amount of equipment that can be used will depend on the budget of the broadcast and availability of equipment and personnel.

LOGISTICS AND FORMAT

Preparation is the key to a successful broadcast. Planning for every eventuality may avoid critical problems before they occur during a broadcast. Not every contingency can be worked out in advance. The producer/director must be flexible and creative enough to smooth over any problem situations. With several exceptions, such as a total power outage, pregame planning can save a broadcast.

The producer must deal with any number of problems and details. Arrangements for crew passes, security, meals, lodging, parking, and other details must be handled. The producer must work with the public relations and sports information personnel for both teams. The producer must handle the public relations for his station. He represents the station or network to others and must be aware of it at all times. The producer must have patience, creativity, and cunning to keep everything and everyone working as a team. Many egos are involved and must be handled with care.

Once the basic planning and logistics are handled, the producer and director plan for the actual game broadcast. The philosophy or approach to the game must be determined. After researching the teams and players, possible story lines are picked out. Once they are identified, efforts are made to provide support. Appropriate video cuts and scripting are developed. The entire game is formatted and planned to the final detail. They must know when commercial breaks will run and what feature material or interviews should be used. It is vital to the look of the broadcast that all prepared materials have excellent production values.

Any graphics that are to be used in the broadcast must be readied. The style, type, color, and format must be picked and loaded into the character generator. All the statistical information must be loaded ahead of time. The players' names, hometowns, and other information must be typed and formatted. Material such as photos can be loaded into the still storage unit, to be called up as needed. The more material that can be loaded and checked, the better.

Contingency planning continues through the pregame activities. Each producer and director has her own idea on what areas will be of primary concern. Play the "what happens if?" game. What happens if one team leaves the court before the game is over? What happens if the communications system for the director and crew won't work?

CAMERA PLACEMENT

The cameras are the eyes of the viewers. Camera placement and use are critical to the look of the game. The placement of cameras varies with each sport, depending on the nature of the sport and availability of camera locations. In football, it is common to have three cameras on the top and along the sides of the field. Cameras are placed on the 35-yard lines and at midfield. When the line of scrimmage moves in front of each camera, it may become the main action camera. Once the main action camera is selected, the other cameras receive specific assignments. Camera One may be the action camera, with Camera Two covering the receivers on the far side of the field, with Camera Three covering the near side receivers. An endzone camera is perfect for covering field goal or extra point attempts. It's good for isolating selected players. It also provides a look down the sidelines that is great for checking if a player stepped out of bounds.

In basketball coverage, it is common for two cameras to be up on the side, and one or two cameras down at the court level. One of the two up cameras is designated the main action camera. The second up camera is used for hero shots (a view of the player who blocked a shot, made a basket, etc.) and other specialized views of the game. The floor camera is at one end of the court and may be placed under the basket or shoot from the corner. The low shot gives a dramatic look that is excellent for a replay. The director and camera people must be careful not to violate the 180-degree rule. If you cut from one end camera to another end camera, people may be confused as to the direction of the action.

Baseball is the hardest of the three sports to cover on television. The field is large, the ball is small, and much of the action takes place *away* from the ball. A camera is usually placed "home-high," which is behind home plate and up above the field. Two other common positions are "first base-high" and in center field. If a fourth camera is used, it may be placed at "third base-low" or "first base-low." Knowledgeable camera people are vital when covering baseball. They must know the game well enough to anticipate and work the game action. A good director will create a list of shots to work from. It will list every situation and what camera is responsible for what coverage. It takes careful planning and practice.

Once the camera angles and shots have been tried and checked out, the director makes sure the replay machines and operators are ready to roll. Football is perfectly designed for use of the instant replay. There is a pause at the end of each play that allows the director to show the play again, possibly several times, from different angles. The director chooses which replays will be shown, for how long, and at what speed. Basketball replays are limited due to the flow nature of the sport. Big plays can be saved and shown as time permits. Baseball offers plenty of time for replays due to the stop-and-go nature of the sport. Each replay operator signals or tells the director they have a replay ready and what it might be. The director picks the replay to be used. The replay is commonly recorded on 1-inch videotape machines. The operators record a video feed, usually direct from a camera, until the play is complete. The operator quickly rewinds to the beginning of the play and starts the tape moving when the director calls for it. A tone can be inserted on the tape as a starting point for use by the operator.

GAME DAY

On game day, everyone involved shows up to run a facilities check, communications check, camera practice, and a production meeting to smooth out any problems and make sure everything and everyone is ready. After the meeting, the producer/director heads to the truck to make all final preparations and handle last-minute problems. The producer/director confers with the chief engineer, who makes sure any technical problems are solved and things are ready to roll.

When it's time to begin, the director calls for specific recorded materials and live camera shots to begin the game. There is usually a prerecorded open that will be run from the main studio. The producer helps the director work for a specific look and story line for the broadcast. The technical director listens for the commands of the director and pushes the appropriate button on the video switcher to put the desired camera shot or replay on the screen for the audience. There will be floor director who is on the court or field to make sure things operate on the television time schedule. When a time-out is called by a team, a commercial is run from the TV studio. Action is not allowed to resume until the floor director signals that TV is back from the commercial break and ready to go. The floor director works with the officials and public address announcer on when the players will be introduced, when the teams come out on the court or floor, when the band should play, and when action should begin.

A good director learns to anticipate, to look ahead to the next monitor to see what's coming up. When a basketball player steals the ball and goes the length of the court to slam it in, the director will anticipate the follow-up replay and how it will be shown to the audience. As the director calls for the replay over the intercom, the talent also hears the message and responds with the appropriate audio. The director is always looking for the next shot. A director must lead and motivate his crew to get their best efforts and a professional looking broadcast.

The announcers should supplement the camera work and call attention to an excellent shot or angle or coverage. Since many of the shots are already formatted, the announcer can anticipate how the audience will be seeing that particular play and can describe it accordingly. Television coverage is a team effort and the crew deserves to have their efforts recognized.

Of the three major sports, basketball is the easiest to do on television. It takes place indoors where there are no weather problems and the lighting is usually constant. The area to cover is relatively small and only ten players are involved in the action. It is easy to get in close and personal with the team members and the action taking place.

Football is played in all types of weather and in a large area with constantly changing lights and shadows. The players are hard to identify, due to their uniforms with helmets, and the fact that 22 people are involved in each play. The sport is well suited for the use of the replay with time for analysis of the plays and plenty of breaks in the action for commercials. The hard-hitting contact comes across to the viewer in a very exciting way. The control of the replay speed adds grace and power to the action, allowing the audience to understand and enjoy the situation.

Baseball is probably the hardest of the three sports to do. There is a lag in the action, and then the action is divided and hard to cover when it does occur. It is the sport where the action is not just where the ball is. It takes place away from the ball, so several areas must be covered to give a complete picture of what is happening and still get in close enough to the action. Because there are so many situations and possible strategies that occur in baseball, the director has a hard time formatting the shots. The camera people must be on their toes and ready to switch shots according to what is happening.

The magic of television and the equipment and techniques that produce the magic are constantly changing. Every year the sports viewing audience gets a different view, or perspective, of the sporting events. The difference between being at the game and watching it on television is quite apparent. The game or event, as viewed on TV, is one of constantly changing angles and views at different speeds. Television takes the play or action and gives it to the audience over and over again. From "skycam" to special computer graphics and designs, television is constantly trying to take the same looking events and make them new and different to the audience to keep their attention.

17

Electronic News Gathering

Video highlights and voice bites are an important part of any newscast on television. Today, most television stations assign their news photographers to cover many of the sporting events within their viewing area. These sports include high school, college, amateur, and professional sports.

GATHERING VIDEO

There are two ways of gathering video for the sportscast. The most common way is to send a photographer to the event with one camera and one recorder to videotape all or part of the game. Then the photographer returns the tape to the station where it is edited for the newscast. The second and least common way of sports gathering is to cover the event live, using microwave or satellite technology. Microwave technology is the simplest form of transmission, but is limited to a distance of about 35 miles, depending on the obstructions and the terrain. When microwave technology cannot be used, it may become necessary to use a communications satellite. This is a more costly and involved process. The station must purchase or lease an uplink truck. The uplink truck has a portable satellite dish and transmitter and sends the audio and video signal to the satellite that is 22,300 miles above the earth's equator. The station then leases time on a transponder on the satellite from the owners. The lease may be for as little as 5 minutes to as long as necessary. The signal is then received at the downlink satellite dish, usually located at the television studio or at the station transmitter site.

CREDENTIALS

Permission must always be obtained before entering any sporting event. There are usually some restrictions on the photographers, limiting them to certain areas from which they may shoot. The management and governing boards may limit the station to only 1½-2 minutes of video footage. Further restrictions usually prevent TV stations from airing any highlights until the sporting event is over. If you are airing highlights from another station, you must have permission from all responsible parties. Most stations will require a courtesy key, i.e., "Courtesy WISH-TV" or "Courtesy CBS Sports."

(This material is provided by Professor Jay D. Burks, Vincennes University)

Sports directors, news directors, or assignment editors should send all requests for press passes and credentials in writing at least two weeks in advance. Larger sports events such as the Indianapolis 500, or many professional sporting events, require that all requests for credentials be sent at least 30 days prior to the event. It is always a good idea to make your request early as there is usually a limit on the number of credentials to be issued. For some events, it may be possible to obtain a season pass if your station is carrying a majority of the events. A number of the credentials, such as those for motor sports racing, require all individuals requesting credentials to sign a waiver releasing the promoters and track owners of all responsibility should you receive an injury at their facility.

ENG EQUIPMENT

Electronic News Gathering (ENG) usually involves using one portable camera and one portable video recorder to capture all the big plays and reactions to those plays. ENG photographers must understand their equipment and know how to operate it properly plus do some simple trouble-shooting. The following is a brief inventory list of equipment and accessories necessary for an ENG shoot.

Portable color camera with a fresh battery
Portable video cassette recorder with a fresh battery
14-pin camera cable or coax cable with BNC connector
Extra batteries for the camera and recorder
Several rewound video cassettes with record pins
Portable battery-powered light (called a Sun Gun)
Microphone (usually super cardiod and cardiod)
Microphone cable
Tripod and adaptor plate (optional)
Two-way radio

Before leaving the television station, *always* set up all the portable equipment and make sure it makes good audio and video. If there is a problem, have it checked out by a broadcast maintenance engineer before traveling to the game.

Video Formats

There are several videotape formats being used by broadcasters for news gathering. The Beta and MII formats have the best overall quality of the portable formats. These two formats utilize a ½-inch video cassette that usually has a tape length of about 20 minutes. The ¾-inch U-Matic format is the oldest format still currently used for ENG, and its picture quality is only average at best. It is a reliable and relatively inexpensive format. Most stations still have some ¾-inch machines at their studios. The biggest disadvantage of the U-Matic format is the problem of a two-piece system. The camera and video recorder must be connected by a cable. The cable is one of the accessories that breaks down from time to time. The newest format for stations desiring to use a low cost alternative is the S-VHS or Super VHS format. All formats, except the ¾-inch format, are available in a one-piece camera recorder unit called a *camcorder*. The use of the camcorder in the various formats, such as Beta, MII and S-VHS, eliminates the need for a 14-pin camera cable or a coax cable with a BNC connector.

Camera and Filter Setups

Setup procedures for the camera and recorders are as important as the footage that is shot. Most cameras have similar controls for setup and they may be labeled a little differently, but they all perform essentially the same functions.

Each camera is different when it comes to switches necessary to turn the actual camera on. The operator should consult the operator's manual for the correct procedure. Once the camera is powered up, the photographer should record about 30 seconds of color bars on the beginning of each tape that is used. This is necessary to adjust the video levels, chroma levels, and the color itself when playing back a videotape on the air through a Time Base Corrector (TBC). After recording the color bars, turn the bars off. The next adjustment is to select the proper degree Kelvin filter setting for the existing lighting conditions. 3200-degree Kelvin is the proper setting for studio lights, portable lights, and most artificial lighting sources. 5000-degree Kelvin (K) to 6000-degree K is the setting used for outdoor shooting with natural sunlight. The only exception might be sunrise or sunset when the 3200-degree K would be the correct choice for those conditions. The filter wheel will probably include one or two positions labeled cap or close. These positions prevent all light from reaching the pickup tubes or Charged Couple Devices (CCD). This position should always be used when the camera is not in use to prevent damage to the internal optical system.

Once the degree Kelvin filter has been selected, the black balance is next. Usually the operator holds the switch or pushes the button to perform this function. The viewfinder will usually have some form of an indicator to show that this procedure is complete. The black balance makes sure the camera sees black accurately. For example, if the black levels were too high, black might look like charcoal gray or lighter. If the black levels are too low, dark colors, such as navy blue, would look like black.

After completing the black level, the photographer must do a white balance. The white balance is extremely important to make sure the colors are accurately reproduced. Most photographers will put the camera on automatic iris. This means the camera will automatically adjust the f-stops, based on the lighting levels. To do a white balance, the operator must find something pure white to shoot, such as a towel, paper, a car, etc. After filling the majority of the screen with white, push the white balance button and the colors should be accurately reproduced on the videotape. It is a good idea to do a white balance every time the lighting conditions change, or when you power down the camera, or when changing batteries. Although many cameras have an internal battery for white balance, it is still a good idea to redo white balance periodically. It should be noted that all viewfinders on portable cameras are black and white, so the photographer has no way of knowing if the color is correct.

The final setup procedure is to align the pickup tubes. When the tubes are not aligned properly, the image will be fuzzy and may have red or blue halos around all the objects. To do an alignment, often called *auto centering*, the photographer locates a horizontal and vertical intersecting line, such as a window pane. Focus sharply on these intersecting lines and hold the auto centering button for a couple of seconds.

Be sure you don't keystone the picture when you shoot the lines. Keystoning occurs when the camera is not shooting the object straight on or, in other words, at a 90-degree angle. Keystoning is shooting any object that has little depth at any angle other than 90 degrees. This causes one side of the object to appear smaller than the other side. Shooting the lines straight on a 90-degree angle will prevent any distortions.

CCD cameras utilize a computer ship to replace the pickup tubes. The pickup tube converts light waves into a video signal. Pickup tubes have a tendency to retain an image on the surface of the tubes for a brief period of time called *lag*, *sneak*, or *comet tailing*. If the image remains on the tube, it's called a *burn-in*. Burn-ins are caused by bright lights, the sun, or bright reflections while holding a static shot too long. CCD cameras do not have any tubes and therefore do not retain any images. These cameras can be pointed at bright lights without any problems. The CCD camera does not have alignment problems like cameras with the pickup tubes. Bumping a camera can cause tubes to be out of alignment. This may cause the pictures to have a blue or red halo effect around all objects. CCD cameras do not require alignment in the field.

The camera is now properly set up and you're ready to begin shooting. Check the lighting levels first. If the levels are too low (there should be a character generator or waveform monitor) in the viewfinder to evaluate the lighting levels, you can add artificial lights or add video gain. Video gain is electronically amplifying a weak video signal. High-quality cameras have gain of usually 6 decibels, 12 db, 18 db and sometimes 24 db. The higher the db, the greater the amplification of the video signal. Usually the higher the db, the more grainy the picture and resolution (sharpness). Low end, less expensive cameras will have more graininess in the picture when video gain is added.

RECORDING CATEGORIES

ENG photographers are routinely called upon to record the following categories of audio and video:

Actuality: This is an interview on-camera with an athlete, coach, expert, or witness to a sports story. The shot is usually a medium close-up or close-up shot.

B-roll or **Cut-in** (insert): This is a video footage of the main subject area, i.e., the game itself. Always shoot natural sound (environmental sound) with all the B-roll that is shot for each story or sporting event. The photographer should shoot a variety of establishing location shots and a lot of medium and close-up shots.

Reaction Shots (cutaway): These shots are reactions to the game or event. Examples include crowd shots, shots of the scoreboard, cheerleaders, coaches, sideline player reactions, and so on. These are used, when editing, as separators and to prevent jump cuts. Jump cuts are two similar shots edited back to back. This should always be avoided. Always shoot natural sound when shooting reaction shots.

Narration: A sportscaster or reporter will record a story or summary of the game and provide segues for the voice bites for the completed story. This is an audio-only recording and is usually done at the station, but is sometimes done on location.

SHOT COMPOSITION AND FOCUS

Shot composition is an extremely important element of photography. When shooting an interview, it should be shot straight on and centered with either a medium close-up shot or close-up shot. This shot eliminates most of the background distractions. When shooting the subject with a close-up shot, make sure you have enough room for two lines of character generator graphics at the bottom. One line is for the person's name and the second line is for the person's title. When shooting a profile shot, always leave more room in front of the face than behind the head. This technique is referred to as *lead room* or *looking room.* A profile shot is less desirable than a straight-on shot.

The picture must be sharply focused. The proper way to focus an ENG camera is to zoom in all the way on the subject and rotate the focus control on the lens until the image is sharp. When shooting a person, the photographer should zoom all the way in on the eyes and adjust the focus control until the picture is sharply focused. Then the operator may zoom out to the desired shot and it will be in focus, as long as the distance between the camera and the subject does not change. Should the distance change while you're recording, the photographer must *follow focus.* Follow focus refers to adjusting the focus control while the camera is recording the events. The photographer must rely on the image in the viewfinder to determine if the focus is correct. As the photographer zooms in, the image will probably appear to be more out of focus and more focus adjustments will be necessary. If the follow focus is done properly, the image should always appear to be sharply focused as the distance continually changes.

ENG photographers will get a steadier shot if they use a tripod. Many stations have a news department policy of always using a tripod. There are three exceptions to the use of the tripod: when there is no time to set up the tripod to get the shot, when there is no room to set it up, and in sports action coverage. To minimize the shakiness of a handheld camera shot, the photographer should get extremely close to the subject and zoom out as much as possible to get the correct shot. Small focal lengths (zoomed out) shows less shakiness than larger focal lengths (zoomed in). It is not always possible to zoom out for all shots. Usually the operator must zoom in to get the correct shot and the only way to minimize the shakiness of being zoomed in is to continually practice and build up endurance for your upper body and shoulders. Wearing comfortable clothing and shoes can make it a little easier if you're shooting for long periods of time.

Football and basketball are two of the most common sports for colleges and high schools. When shooting footage for these sports, it is important to keep all of the B-roll shots within 180 degrees. That means the photographer must not go beyond the goal posts or the basketball goal, otherwise the players will appear to be running in the opposite directions. When shooting football, the photographer should travel up and down the sidelines following the line of scrimmage. Be careful to stay far enough out of bounds so there is not interference with the players, coaches, or officials. ENG shots must be held much longer than expected to make it easy to edit the footage into a complete VTR sports package.

To start shooting football footage, begin with a shot of the players in the huddle. As the huddle breaks, zoom out and show the offensive and defensive players at the line of scrimmage. This shot should include the quarterback. As soon as the quarterback begins calling the signals, begin zooming in on the quarterback. Once the quarterback has the ball, the photographer must follow the ball. If the ball is handed off for a run, zoom in for a medium shot and follow the run. Remember to frame the shot with adequate headroom. Once the run is complete, get a close-up or medium close-up of the key offensive and defensive players on that play. This is sometimes referred to as a *hero shot*.

Make sure all shots are held long enough for editing. When there is a pass play, follow the ball in the air and, as the ball is caught by the receiver, zoom in some and follow the run until it is complete, then follow with a hero shot. Other plays, such as a punt, kickoff, or field goal, involve zooming in on the kicker at the beginning, then zooming out to a wide shot for the kick, and then following the ball once it is in the air. When there are dead periods of action, such as a time-out or end of a quarter, use the opportunity to shoot reaction shots. Always shoot a large variety of shots and hold them for about 20 seconds each. The photographer can also get nice shots from the top of the press box or the top of the bleachers. This "aerial" view will show more of the activity on the field. One bit of caution: when shooting night football, there is a concern about getting a burn-in on the pickup tubes in the camera. The photographer must be careful about having a static shot that includes the bright lights or a bright reflection.

Basketball highlights are easier to shoot. The most important thing to remember is to always keep the ball in the shot, except when shooting reactions shots. To begin with, get a medium shot of the players ready to throw the ball inbounds and then begin to zoom out. Once the ball is put in bounds, stay with the medium-long shot of the players bringing the ball upcourt. As the players cross the mid court line, zoom out to show the floor from the goal to the time line. This will ensure that the ball is always in the shot while it is passed around. As the ball approaches the goal, zoom in a little to follow it through the hoop or follow the rebounding activity. When there is a foul shot, start out with a tight shot of the shooter, then zoom out before the ball is released and follow the ball to the hoop. As in football, basketball has the hero shot. These shots will include the player scoring the basket, committing fouls, shot blockers, and other play makers. A variety of reaction shots are necessary for editing. Always hold the shot longer than usual for editing purposes.

VIDEO HIGHLIGHTS

View the sports highlights of local television stations and see how they shoot various games. It takes practice and endurance to do a quality shoot. Some stations will send an assistant with you to these shoots. This person, often referred to as a spotter, should make sure that no one is injured, or equipment damage, if the play suddenly travels out of bounds. Don't forget to shoot natural sound during the game. A shotgun microphone does an excellent job in this situation. When recording sound, make sure you are not talking with your spotter or someone else or it will be recorded on your tape. When shooting highlights, remember to have some videotape and battery power left to get an interview or two after the game. By that time, the photographer is somewhat fatigued and a tripod is recommended. The portable light should be used for the interviews on game night.

It is a good idea to label the tapes with masking tape to indicate the game and suggested order of shots for that game. If you are shooting more than one game on the same day, do *not* record two different games on the same tape. This can be extremely confusing to the person editing the video footage.

Appendix A

Forms

FOOTBALL SUMMARY FORM

Date:

T/QTR		SCORE
	TD:	
	PAT:	
	TD:	
	PAT:	
	TD:	
	PAT:	
	TD:	
	PAT:	
	TD:	
	PAT:	
	TD:	
	PAT:	
	TD:	
	PAT:	
	TD:	
	PAT:	
	TD:	

T/QTR		SCORE
	TD:	
	PAT:	
	TD:	
	PAT:	
	TD:	
	PAT:	
	TD:	
	PAT:	
	TD:	
	PAT:	
	TD:	
	PAT:	
	TD:	
	PAT:	
	TD:	
	PAT:	
	TD:	

1 2 3 4 5 6 7 8 9 10 11 12 13 14 15 16 17 18 19 20 21 22	**FIRST DOWNS**	1 2 3 4 5 6 7 8 9 10 11 12 13 14 15 16 17 18 19 20 21 22
1 2 3 4 5 6 7 8 9 10 11 12	**3RD DOWN CONVS**	1 2 3 4 5 6 7 8 9 10 11 12
1 2 3 4 5 6 7 8	**TURNOVERS** **FUMBLES**	1 2 3 4 5 6 7 8
1 2 3 4 5 6 7 8	**INTERCEPTIONS**	1 2 3 4 5 6 7 8
1 2 3 4 — 1 2 3 4	**TIME OUTS**	1 2 3 4 — 1 2 3 4

PENALTIES

1	2	3	4	5	6	7	8	9	10

1	2	3	4	5	6	7	8	9	10

ATTENDANCE: **WEATHER:** **NEXT BROADCAST:**

Basketball Form

TEAM: ______________________

Shots:

1	2	3	4	5	6	7	8	9	10	11	12	13	14	15	16	17	18	19	20	21	22	23	24	25	26	27	28	29	30	31	32	33	34	35	36	37	38	39	40	41	42	43	44	45	46	47	48	49	50
51	52	53	54	55	56	57	58	59	60	61	62	63	64	65	66	67	68	69	70	71	72	73	74	75	76	77	78	79	80	81	82	83	84	85	86	87	88	89	90	91	92	93	94	95	96	97	98	99	

Turnovers:

1	2	3	4	5	6	7	8	9	10	11	12	13	14	15	16	17	18	19	20	21	22	23	24	25	26	27	28	29	30	31	32	33	34	35	36	37	38	39	40	41	42	43	44	45	46	47	48	49	50

1	2	3	4	5
6	7	8	9	10
11	12	13	14	15
16	17	18	19	20
21	22	23	24	25

NAME	#	1st	2nd	FOULS	3rd	4th
				1 2 3 4 5		
				1 2 3 4 5		
				1 2 3 4 5		
				1 2 3 4 5		
				1 2 3 4 5		
				1 2 3 4 5		
				1 2 3 4 5		
				1 2 3 4 5		
				1 2 3 4 5		
				1 2 3 4 5		
				1 2 3 4 5		
				1 2 3 4 5		
				1 2 3 4 5		
				1 2 3 4 5		
				1 2 3 4 5		

Team Fouls: 1 2 3 4 5   1 2 3 4 5   1 2 3 4 5   1 2 3 4 5

Football Form

Record: ______________ Team: __________________________ Coach: ______________________

PUNTER: KICK-OFF: FIELD GOAL:			1ST DOWNS
			PASS ATTEMPTS
NUMBER OF PLAYS PER SERIES FROM START			COMPLETIONS
			INTERCEPTIONS
			FUMBLES
			PENALTIES

TEAM ____________ **vs.** ____________ **At:** ____________ **Date:** ________

PLAYERS	1	2	3	4	5	6	7	8	9	10	11	12
 Sub												
 Sub												
 Sub												
 Sub												
 Sub												
 Sub												
 Sub												
 Sub												
 Sub												
 Sub												
SUMMARY												

K ____________

HITS ____________

WALK ____________

LOB ____________

ERR. ____________

SB ____________

	1	2	3	4	5	6	7	8	9	F

WP ____________

LP ____________

LINE SCORE:

____________ __ __ __

____________ __ __ __

Appendix B

Selected Sportscasting Terms

Actuality The pictures or sounds of a newsmaker or news event.

B-roll (Cut-ins) Video footage of a main subject area with natural sound.

Camcorder A one-piece video recorder-camera unit.

CCD Camera Charged Couple Devices—utilize a computer chip to replace video pickup tubes in a camera.

Chroma-key A process that allows the picture from one camera to be keyed or superimposed in a portion of the picture from another camera or source.

Close-ended Question A question with a specific set of answers to choose from.

Courtesy Key Graphics run over video taken from another TV station or network, giving them credit as the source.

Cume The number of different (counted only once) people who watch or listen to a station over several time periods or dayparts.

Cutaway When a TV reporter is shown listening or taking notes during or after an Interview to show he was participating.

Dedicated Line (Loop) A telephone line, preconnected by the telephone company. Provides a balanced, direct line from the game site to the studio for a broadcast.

Depth of Field Refers to the portion of an image in front of a lens that is sharply in focus.

Director The person who makes decisions about the actual TV broadcast, with regard to camera shots, number and length of replays, and so forth.

ENG Electronic News Gathering—usually involving a portable videotape camera and recorder.

Flagship Broadcast station that originates the signal for the rest of the sports network.

Frequency Extender Takes lower-end frequencies of sound that are usually left out of a telephone transmission, mixes them into the telephone line to be decoded and used at the receiving end. Gives the sound more depth and quality.

Lavaliere Small unobtrusive microphone that is clipped on clothing or hangs around a person's neck. Used mainly by studio anchors.

Lead Scripted information to be read prior to an actuality, used to set the scene.

Microwave System A type of delivery system for local TV transmissions, usually from a remote truck to the station's microwave antenna.

Mixer Portable audio control board to control different audio inputs, combining them into a usable signal.

Moiré Effect A wavy, shimmering picture caused by fine, high-contrast patterns, such as clothing with narrow stripes or small checks, that TV cameras cannot handle.

Narration When a TV reporter records an audio-only story or summary of the game to go with the video plus segues for interviews.

News Release Information put into a specific form for the media to use as news.

Open-ended Question Questions that don't require a specific type of answer. Subject must use their own words and thoughts to complete their response.

Parabolic Microphone An omnidirectional microphone faces into a parabolic reflector (which is concave) that concentrates desired sound waves from a specific source.

Producer Person in overall control of the broadcast in radio or television.

Ratings The number of people who watch or listen, shown as a percentage, of the total possible listening or viewing audience.

Reaction Shot Video shots of reactions by the crowd, players, coaches, and others, to the game or event.

Reverse Question A question asked during an interview is asked by the reporter again, but with the camera in a reverse angle to show who asked the question.

Ridiculous Association A memory technique involving odd substitutions or exaggerations, and then associating or relating them to names.

Rights Fees Money paid by a broadcast station to gain exclusive rights to carry a school or team game.

ROSR (pronounced RO-zer) Stands for Radio On-Scene Report. Usually a reporter's monologue describing the scene or action as it occurs.

RTU or RPU Remote Transmitter Unit or Radio, Portable Unit—portable radio transmitters used to send a signal from a remote site back to the studio for broadcast.

Screener A person who takes incoming telephone calls and stops questionable or objectionable callers from getting through to the talk show host and on the air.

Shared Feed Announcers from one TV station go on the air while the second station cuts away to commercials or other materials. The process is then reversed for the other station. The same game video is used for both stations, but with individual audio.

Shotgun Microphone Cancels sound in all directions except for a narrow angle at the front, thereby cutting out extra noise and allowing concentration from a desired sound area.

SID Sports Information Director—Person responsible for media relations for a team.

Site Survey A TV station representative examines the physical layout of a stadium or arena to access equipment and other needs for a TV broadcast from that location.

Split Game When two TV stations carry the same game using two separate signals or outputs. The basic video action is the same for both stations except at predetermined times when a separate camera is used by one station. Audio is always separate.

Splitter A connector that takes a single signal and splits that signal into two or more outputs.

Spotter A person to help announcers locate and identify players. usually used in football.

Stand-up A field TV reporter appearing on-camera.

Tag Scripted information read after an actuality to clarify and finish the story.

Tape Delay A game recorded on audio- or videotape to be played back to the audience at a later time.

Telephone Interface Equipment that allows a telephone call to run directly through the audio mixing board, allowing better quality, easier access and use.

Transponder Small transmitter/receiver units on a communications satellite that receives the signal and rebroadcasts it back to earth.

Trial Balloon Information given to a reporter to test public reaction to an idea.

Uplink Equipment used to send a signal to a communications satellite for rebroadcast.

Voicer A recorded audio report from a reporter at the scene that is usually complete in itself.

Wrap A type of audio cut that has an open and close report wrapped around an actuality.

Appendix C

Organizations

American Sportscasters Association, 5 Beekman Street, New York, NY 10038

Associated Press, 50 Rockefeller Plaza, New York, NY 10020

Broadcast Promotion & Marketing Executives, 6255 Sunset Blvd., Suite 624, Los Angeles, CA 90028

College Sports Information Directors Association, c/o Mr. Fred Nuesch, P.O. Box 114, Texas A & I University, Kingsville, TX 78363

Radio-Television News Directors Association, 1717 K Street NW, Suite 615, Washington, DC 20006

United Press International, 360 N. Michigan Ave., Chicago, IL 60601